Fearless Spirituality:

What Sages Knew
and
Science Discovered

Also by Lee and Steven Hager

*The Beginning of Fearlessness:
Quantum Prodigal Son*

*The Gospel of Thomas: Where
Science Meets Spirituality*

*Religious or Spiritual:
How the Difference Can Affect
Your Happiness*

Available in print editions and a wide
variety of eBook formats

Lee and Steven Hager

Fearless Spirituality:

What the Sages Knew
and
Science Discovered

**Oroborus
Books**

Fearless Spirituality:
What Sages Knew
and
Science Discovered

Copyright © 2013 by Lee and Steven Hager

Published by **Oroborus Books**

The Beginning of Fearlessness/Oroborus Books
website and blog:

http://thebeginningoffearlessness.com

ISBN: 978-0-9785261-8-4

LCCN: 2013907896

This book is dedicated to everyone who feels certain there must be something greater than what the eye sees, the ear hears or the hand touches.

I hold it true that pure thought can grasp reality as the ancients claimed—Einstein

Since before time and space were, the Tao is. It is beyond is and is not. How do I know this is true? I look inside myself and see—Lao Tzu

CONTENTS

Fearless Spirituality:

What Sages Knew
and
Science Discovered

Introduction

What is 'fearless spirituality' and why should you care? If you're reading these words, it's probably safe to assume that you are interested in spirituality, but you may also feel something is missing. That's where fearless spirituality comes in. You were not meant to seek continually, but to find. You were not meant to keep wondering, you were meant to understand. You were not meant to feel disconnected from the Divine, but to enjoy a direct, personal relationship. You were not meant to live in fear, but to rest in complete peace and security. Does that sound impossible? If so, than why bother with a spiritual quest? Why not stick with what the world has to offer?

Secular history repeatedly demonstrates one fact: most people believe that fearlessness comes from the security attained with money, power and control. But this same history also makes it clear that no amount of money, power or control has ever been enough to insulate anyone from misery, illness, pain or death.

Humanity has already tried virtually every combination of political, social and economic system available; history has witnessed the eventual failure of each one. Yes, scientific and technological advancements have improved many aspects of life. Nonetheless, war, starvation, disease, poverty, ecological and natural disasters continue to ravage our planet and leave us in fear. And many who have turned to religion have discovered the 'fear of God' rather than the solace they were seeking.

As Alan Watts pointed out over sixty years ago, "In the best of times 'security' has never been more than temporary...but it has been possible to make the insecurity of human life supportable by belief in unchanging things beyond the reach of calamity." But such 'spiritual comfort' has been attacked by modern science, which has often delighted in reducing religious beliefs to the level of fairytales or claiming that the concept of God is both superstitious and logically unnecessary. As Watts went on to point out, "Yet for all [science] has done to improve the conditions of life...the price of their miracles in this world has been the disappearance of the world-to-come...logic, intelligence and reason are satisfied, but the heart goes hungry."

In an effort to feed a heart hungry for the Divine, some continue to cling to religion, but many others have abandoned religion, disappointed. In order to draw straying sheep back to the churches, some religions have turned to a message of abundance, promising that God wants believers to be rich. The New Age 'law of attraction' also encourages the belief that the universe is little more than a vending machine ready to fulfill our material desires. Although the appeal of financial security may appear to work in the short term, the long term will inevitably demonstrate that it's impossible for money or things to eliminate our fears. Currently there's an upsurge in spirituality, a personal quest for something greater than the material. Unfortunately, many seekers have no idea how to go about their quest and end up engendering more questions than answers. As Jesus said, "You will know the truth and the truth will set you free." (John 8: 32) Truth is the foundation of spiritual fearlessness, but if we don't know how to find truth or identify it when we see it, fearlessness is impossible.

Our situation would appear bleak indeed if it were not for another history, one that's been quietly written throughout the centuries by thousands of men and women. They each lived a life of fearlessness regardless of how chaotic the world appeared to be. How? By personally *experiencing* what lies beyond the material. This first-hand experience is called *gnosis*, a Greek word that means 'knowing' through direct experience rather than intellectual learning. Instead of finding security in this constantly changing world, their fearlessness was based on a rock solid foundation of changeless truths that flow from Source.

Fear originates in misperception. Fearlessness is the result of communing directly with the Divine; seeing the universe as it is, not as we think it is. Direct experience is *not* a special blessing held out to a privileged, chosen few. On the contrary, spiritual truth calls to all of us, but few choose to listen and fewer still choose to answer. In fact, each of the sages we'll be discussing discovered that the treasure they were seeking was 'concealed' in plain sight. The 'knowing' that will allow you to enjoy their same level of security and spiritual fearlessness is something that you already possess!

The spiritual adventurers who have gone before us have not only left us with their wisdom, their words are currently finding support in the field of quantum physics. Although many religious teachings have been unable to stand up to scientific scrutiny, you'll find that the words of the fearless sages we'll be examining are strengthened by it. Don't worry; you don't need a background in science to understand the research we'll be discussing. We're not scientists either, so we aim for the most straightforward explanations possible.

If your own spiritual journey has felt confusing, frustrating or you feel that there are still some important pieces missing from the puzzle; there is no need to despair. Why do some spiritual seekers reach spiritual fearlessness while so many others continue to bounce from one idea to another? Why do some remain fearful even though they've gathered mountains of information and spent untold hours in spiritual practice? The answers to these essential questions are the focus of this book. After all, if we remain fearful and confused, what's the point of seeking? There are no secrets and nothing magical about what we'll be discussing, yet the answers are astonishing and have the potential to dramatically change your life. (We say 'potential' only because the outcome is up to you.)

On occasion, we'll draw from writings that various religions consider 'sacred,' but more often we'll be quoting the wisdom of a wide variety of fearless spiritual sages, from ancient to modern. Why? The 20th century sage, Eknath Easwaran, pointed out, "The spiritual teacher must know every inch of the way, every danger and pitfall, and not from books or maps or hearsay. He must have traveled it himself, from the foothills to the higher peaks. And he must have managed to get back down again, to be able to relate to students with humility and compassion." Many so-called 'holy books' are based on the hearsay of followers rather than the firsthand experience of the successful seekers they followed. This book would not be worthy of your time or attention if it were not based on the direct wisdom of those who have successfully walked the road you also want to tread. As the Sufi sage Rumi said, "When setting out on a journey, do not seek advice from those who have never left home."

We'll be drawing on our own experiences as well, not as a means of establishing our 'spiritual credentials,' but rather to demonstrate that the answers we all seek, and the spiritual fearlessness that results, are well within the reach of each and every one of us. Our search was long and at times challenging, but we realize that by sharing our many misadventures, we may well help you avoid some of the same pitfalls and long detours that we took.

In the gnostic *Gospel of Thomas*, Jesus said, "I shall give you what no eye has seen, what no ear has heard, what no hand has touched, what has not arisen in the human heart." Are you willing? If so, Jesus continued, "Let one who seeks not stop seeking until one finds. When one finds, one will be astonished, and having been astonished one will reign, and having reigned, one will rest." We all know that it's impossible to rest when we're afraid. What Jesus offered was not only astonishment at what we would learn, but the fearlessness that can come only when we're absolutely secure within what we know to be true. We found these words to be true, and we know that you can too. It is our greatest joy to share with you the wisdom of successful seekers. May you choose to hear the call and join your heart to theirs in fearless spirituality.

You've been walking the ocean's edge, holding up your robes to keep them dry. You must dive naked under, and deeper under, a thousand times deeper!—Rumi

Part One

The Ultimate Obstacle

I. The Perception Trap

All the veils are one veil. Other than that one, there is no veil. That veil is this existence— Shams-iTabrizi

Although our goal is the priceless treasure of spiritual fearlessness, it is not something we can grasp unless we first clear away the rubble that obscures it from our vision. We might imagine that we'll be faced with removing tons of debris, but there is actually only one obstacle all spiritual seekers face. This obstacle is simple, yet difficult to recognize until we know what we're looking at. Once we do learn to recognize this foundation obstacle, everything that's been built upon it will become clear, finally allowing us to see with new eyes. Strange as it might seem, this foundation is made up of our own misperceptions. These misperceptions are so ingrained, so much a part of who and what we feel certain we are, they've become virtually undetectable. And as you can imagine, it can be challenging to deal with something that's ubiquitous yet nearly invisible, but it's far from impossible.

How do we become entangled in misperception? The moment we're born, everyone and everything around us plays a part in the conditioning that builds our self-understanding and eventually our world view. As our senses constantly draw in new information, it's rapidly sorted by the brain and then either discarded or filed for future reference. This sorting is based on very specific criteria: since the brain is, first and foremost, a survival mechanism, it puts the highest value on physical safety. As a result, many of the lessons the brain retains are based on fear. Add to this the issue of sense perception. We do not actually experience the world, we are limited to perceiving it through the senses.

Researchers have discovered the eye actually takes in less than half of what it sees, then the brain further distorts the final impression by straining it through the biased information it has already collected. This process is not restricted to vision; the same scenario is true for everything we've ever heard, tasted, smelled or touched. Since the information we collect from the world is first limited by the senses and then by the brain, we cannot know the world, but can only perceive it on a limited basis. What we think of as reality is always subjective and differs from everyone else's perception. In a very literal sense, there are as many worlds as there are perceivers. *Since each of us perceive the material universe differently, objective reality is impossible. As a result, there can be no 'absolute truth' in this world.*

Much of the conditioning the brain receives is subliminal, so we absorb it without noticing. For example, in early infancy we begin to pick up on the emotions of others and use this input to decide which specific people or situations we feel comfortable with and which ones frighten us. We also use this very subtle information to choose how we'll behave around others and gauge our reactions to them. Some of the information our brains gathered in infancy and childhood may no longer be useful as an adult, but the brain will continue to cling to anything it believes is related to our safety. This is why we often retain early childhood fears and emotions long after we've realized the fear is baseless or the emotion is no longer serving us.

On the other hand, much of our conditioning is deliberately taught to us by those closest to us. As soon as we come into this world, they begin to define who we are based on their own limited perceptions and feelings about us. Our strengths, and far too often our weaknesses, are drilled into us as we hear over and over that we are 'pretty, strong, good, bad, talented, bright, stupid, weak, etc. These words, combined with experiences that the brain deems important, begin to form a personality. Our worldview also begins at the feet of those who are closest to us. The musical "South Pacific" illustrated this training with song lyrics that pointed out, "You've got to be taught before it's too late, before you are six or seven or eight, to hate all the people your relatives hate, you've got to be carefully taught." If we were born in another part of the world, we would be raised with an entirely different set of prejudices.

Since we're constantly surrounded by people and events, it's our friends, family, organizations, community, country and even those we may consider enemies, who regularly help shape our worldview. Whether we agree or disagree with them, we're still being subtly influenced. In the internet age, we're also influenced globally as information, ideas and fads sweep the planet in a matter of hours. Without our realizing it, we prove 19th century playwright, Oscar Wilde's observation, "Most people are other people. Their thoughts are someone else's opinions, their lives a mimicry, their passions a quotation." How often do we become interested in something simply because we have the impression that others are interested and we believe we'll be left out if we don't join in?

As 20th century philosopher/mystic Alan Watts noted, "Other people teach us who we are. Their attitudes are the mirror in which we learn to see ourselves, but the mirror is distorted." Unfortunately, there is no way to avoid conditioning since its well under way long before we could possibly be aware it's happening. Even when we become more discerning, the elusive nature of most conditioning remains extremely difficult to detect. Still, we must face the fact that conditioning has an extremely powerful effect. Since the brain is a survival mechanism, it usually opts to find safety in the crowd. Most adopt the traditions, culture, religion and politics of the country or area they live in, rarely, if ever, asking themselves if their way of life is meaningful to them. The brain has taught us all to live on the surface of life and fear its depths. The 15th century alchemist/philosopher Paracelsus realized, "We are fooling away our time with outward and perishing things, and are asleep in regard to that which is real within our self." What can we do to counter the conditioning? We begin by realizing it's there.

Successful spiritual seekers throughout history woke up to the fact that conditioning had put a blindfold over their spiritual vision. Once recognized, they were able to begin letting go of conditioned thoughts. Free of the brain's limited perception, what they saw shocked, amazed and then comforted them. Jesus expressed this beautifully in the gnostic *Gospel of Thomas* when he said, "Let one who seeks not stop seeking until one finds. When one finds, one will be astonished, and having been astonished, one will reign, and having reigned, one will rest." Jesus could say this because he had let go of the conditioning and misperceptions that had clouded his understanding. Please keep in mind that the person who recorded these words in the *Gospel of Thomas* understood Jesus as a fellow seeker, not the god some of his followers eventually turned him into. When the Jesus in *Thomas* said, "Let one who seeks not stop seeking until one finds," he spoke from his own personal experience as a seeker. He wanted others to know that they could have the same experience he had, discover the same truths and reign with him, but it would take some effort on their part. What did he mean by the word reign, and what would he reign over? Understanding what he meant can help us clear up many of our own misperceptions.

Although most of us have been taught that Jesus' earliest followers were a tightly knit group that all shared the same beliefs, there was actually very little consensus among them concerning who Jesus was or what he was teaching. Although some of his followers thought of him as a human sage and teacher, others came to believe Jesus was a God-appointed messiah (anointed one) and warrior king who would destroy their Roman oppressors and establish a theocracy on earth. They not only believed that Jesus would restore freedom and prosperity to the land; they also expected to rule as kings with him. When Jesus was killed, these followers declared that he would triumphantly return to earth to establish God's earthly kingdom within their lifetime, but even these hopes were dashed when Jerusalem was destroyed by the Romans in 70 AD. Instead of asking themselves if they had misunderstood his teachings, many switched their hopes to a heavenly government and supernatural kingship. Although this hope is still alive among

Christians, Jesus had other early followers who understood his use of words like 'reign' and 'conquer' *symbolically* rather than literally. The literal use of these words deepens our misperceptions; the symbolic meaning liberates us from them.

When Jesus told his followers to "be of good cheer" because he had "conquered the world," he wanted them to understand that he was talking about a spiritual experience that's open to all of us. Instead of fighting a physical battle, he meant that he had conquered the conditioning and misperceptions of this world that had once blinded, confused and imprisoned him. Now that he could clearly see himself, the universe, and the Divine as they actually were, he was no longer enslaved to anything in the material realm. Since he now reigned over his past misperceptions, he had figuratively conquered the world and could rest in total security.

Pointing out the reward for letting go of misperception, Jesus said, "You will know the truth, and the truth with make you free." *Since truth can't exist in our subjective world of perception, we must find it apart from the material realm.* To see past what the senses and fearful brain tell us, we must be willing to be liberated from the misperceptions that have tied us to them. This thought was also expressed by the 9th century Indian sage, Shankara, when he said "Who has overcome the world? He who has conquered his own mind." We now know that our first step toward fearless spirituality is letting go of our conditioned misperceptions. Instead of starting with small concepts and working our way up, let's go directly to the root, to the greatest misperception of all.

II. The Connection: Know Yourself, Know Your Source

If you've ever been asked to write something about yourself, you probably included your name, gender, age, work or profession, family and relationships, education, race, nationality, or some of the accomplishments you're most proud of. *Each of these names, labels and titles that we use to describe ourselves are based on the assumption that we are a human being, a body and personality that is separate and distinct from all other humans.* Well of course, you may think, what else would I be? We generally take our humanity for granted; it appears to be so fundamental to our existence; it doesn't even merit our consideration. But is this assumption correct?

The conditioning process that we all undergo is based on two suppositions: first, we are exactly what our brain and senses tell us we are, and second, this world is our reality. But what if these two thoughts serve as the foundation for every misperception we've ever been fed? *What if you are not a body and personality, and never have been? What if the material world that appears to be your reality never has been?* These enormous misperceptions would conceal the amazing truth of who and what you actually are. And, if we aren't aware of our own reality, how could we possibly understand who or what Ultimate Reality is?

Since science has been focused on what can be seen, measured and tested, it has also assumed (until recently) that the body *is* our reality. It is at this point that science and religion have usually parted ways. A few religions and spiritual philosophies teach that we are the body and this life is all there is, but most claim that there is something more beyond the flesh. What that something might be is up for debate, but most religions claim there's an unseen 'spirit world' that exists outside the material world. Is that your reality? You've probably heard this new age maxim: "You're not a body having a spiritual experience; you're a spirit having a human experience." That thought hits somewhat closer to the mark, but spiritual sages who have experienced the truth of who they are have something even more startling to tell us. And surprisingly, quantum research supports what they have to say. But if we're going to understand who and what we are, it's easier to comprehend if we begin with who and what the Source of life is.

III. Our Search for Source

A private detective would become extremely frustrated if you hired them to find someone but gave them either a vague description or one that was totally inaccurate. Nonetheless, that's exactly what we do to ourselves when we try to understand the Divine based on preconceived notions that originated from conditioned misperceptions. As long as we already have an image in mind, the brain's sorting system will keep discarding new ideas and instead looks for confirmation of what it already thinks is true.

The brain builds connections in much the same way as a path is made through the woods; the more it's walked on, the more established it becomes. The more often we return to the same thought, the stronger our neural pathways become and the more likely we are to return to them. When we're in an unfamiliar wilderness area, we usually stick to the well-worn paths unless we have a compelling reason to stray. The same is true of the brain. It considers stasis far more conducive to safety than change. For the brain to make a new pathway, it must be convinced a new thought will be safer than the old thought. Even then, the new thought must be repeated over and over again before it becomes established as the new path. The body also creates chemicals that support our thoughts and emotions and gives us an addictive 'rush' whenever we return to them. This happens even when old thoughts and emotions are no longer serving us. Like any addict, the brain and body crave the accustomed chemical rush until they become addicted to the new chemicals produced by changed thoughts. Obviously, the brain is difficult to convince and harder to change, a helpful concept to keep in mind on the spiritual path!

As we mentioned in the introduction, our own spiritual path was long and at times quiet challenging. Even as children we were both determined to find God. When we met at church and eventually married, we agreed that our spiritual quest would continue to take precedence. We dedicated ourselves to religious service and eventually became quite well-known speakers and teachers. On the outside, we appeared to have unshakable faith, on the inside we felt detached from God and were constantly questioning the church's beliefs. When too many of those questions were asked openly, church leaders felt they were being attacked. They warned us our 'rebellious' behavior would not be tolerated. After a lengthy and intensive self-examination, we resigned from the church, knowing it would mean the loss of family and friends. In retaliation, the church quickly excommunicated us. We spent the next several years exploring other religions, reading sacred texts and looking for secular answers. It would take another book to tell our complete story, but suffice it to say that the harder we searched, the further from truth we felt. Feeling rejected by God, we fell into a deep depression that seemed impossible to escape. Eventually, burdened with financial, family and severe medical problems, we came to the conclusion that suicide was the only way out.

We decided to pray one last time before swallowing the prescription drugs we had hoarded. Our prayers had always been sincere, but that last prayer was nothing like the thousands that had gone before it. Our despair was so overwhelming, it allowed our brains to suddenly do something we had never done before; let go of all personal preferences, attachments, aversions, preconceived notions and conditioning. Instead of once again expecting to have our beliefs confirmed, we told the Divine that we were willing to accept *any* message that was sent to us, no matter what it was. Most important, we meant it! *Our desperation must have also alerted the brain to the life or death situation it was facing, and it cooperated.* We were finally at a point where we could see and accept what Source wanted to show us. Later, we realized that our earlier prayers had made it impossible for us to receive any meaningful answers because what we had actually wanted was validation of the beliefs we already had. Yes, we had always been sincere, but that sincerity was misdirected.

Imagine that you have a bucket you want to fill with fresh water from a sparkling mountain stream, but it's been sitting outside for some time and is now full of muddy, stagnant bug-infested water. Although you'd naturally empty a dirty bucket and scrub it out before using it, we rarely do the same where our thinking is concerned. Instead, we try to pile more information into the brain, even though it's incompatible with the information we've already collected. But the brain is just like that pail, as long as it's full of stagnant information; the fresh won't fit, and it either flows over the top or gets mixed in with the stagnant. We all know that stagnant water will make fresh water dirty, not the other way around. The same is true for our thoughts. *This is a huge issue for the spiritual seeker because the thoughts stored in our brains are usually the muddy, stagnant bug-infested variety.*

Instead of trying to add more misperceptions to the ones we already have, successful spiritual seekers begin by 'emptying and cleaning the bucket' (or at least the willingness to do so). We'll repeat this pertinent fact: *as long as the brain has retained information it still considers safe, very little contradictory information will register.* Before you get too frustrated, it really does makes sense that the brain works this way. Can you imagine how difficult your life would be if you had to collect, retain, remember and sort through each bit of information the brain processes during each moment of the day? The brain's filtering process is both efficient and convenient; without it, we would find it nearly impossible to make a quick decision. Still, its choices do not always serve our best interests.

It was several years after our brush with suicide before we realized how the brain works. Instead of understanding that we needed to let go of our attachments, it took an extreme level of desperation before we were able to surrender our own ideas. However, there is no need for you to get to that level of misery. Instead, you can start to consciously let go of conditioning a bit at a time. As the 13th century Sufi mystic, Rumi, pointed out, "Nothing can teach you if you don't unlearn everything...Be a place unsown, a white paper no writing has stained...and the Merciful One can sow in your blindness the seed of Pure Vision." As far back as the 3rd century BCE, the pagan sage, Plotinus, recognized this necessity too, saying, "It is impossible for one who has in his soul any extraneous image to conceive of the One while that image distracts his attention." To know Source, let's first identify some of the basic concepts that feed our misperceptions.

IV. Outside Gods

Although an extremely wide variety of religions have existed throughout history, they have all believed in either an 'inside' or 'outside' god/s.

What do we mean by that? Let's start by looking at some of the general characteristics that identify each type of god. Outside gods, such as the god of the *Bible*, are separate from their creation. Although they may be described as having human characteristics or pictured with a human-like appearance, they have a very different nature than humans. Their creations are material and mortal, but they are immortal, supernatural spirits. They also separate themselves from their creations by dwelling in an exclusive spiritual realm that exists outside the material universe. Like a king who rules from afar, they either oversee their handiwork from 'heaven' or occasionally visit the earth, but they are under no compulsion to interact with their creation.

The creation stories of outside gods usually follow one of several general patterns. One theme features an 'earth-diver.' In these stories a god sends an animal or lesser being to dive into a primordial soup and find the ingredients necessary to build a world. 'Emergence' stories often feature female deities who birth new worlds from the womb of previous worlds. Some stories feature a 'world parent' that is pulled apart to create the universe. 'Chaos' myths describe a formless, disordered expanse or abyss, from which the god/s draws order. These tales usually also include the belief that order is fleeting and chaos will once again prevail. Some myths are called '*ex nihilo*' because the gods bring the material world forth 'from nothing.' In this scenario, the universe originates from a god's words, breath, dream or pure thought. Christians are familiar with this format since the *Bible*'s creation story in Genesis explains that the material world came into existence as a result of God's thought and the words, "Let there be..."

Often both matter and the life that animates it, is the product of a god's bodily fluids, breath or a tiny portion of their own spirit. But the life that outside gods extend to their material creations is not a permanent gift. They also hold the power to take life away as easily as it's given, quite often through the withdrawal of breath or spirit. Even though these creations may contain a tiny portion of the life-giving spirit of the god, they still have the potential to be either good or evil. This dichotomy leads to myths of cosmic struggles between opposing forces of both good and evil gods and humans. This war between good and evil is a theme that runs throughout the *Bible* (and other holy books) and is used to explain the suffering experienced on earth. According to the *Bible*, humans are being tested to see which side they'll join.

Since outside gods stand apart from their creation and are affected by it only if they choose to be, they can either be interested or disinterested in what goes on in the material realm. Some who believe in outside gods are certain the gods are involved in the smallest happening on earth, such as the outcome of a sporting event. Others believe there is no such thing as free will; that humans are puppets of fate, merely acting out the will of the gods. Those who feel that obedience and worship are necessary are actually demonstrating the belief that it's up to humans to keep outside gods interested in what's happening in the world. To that end, they devise ways to attract the gods such as religions, holy books, rules and moral codes. Since these gods are regularly described as highly temperamental, it comes as no surprise that it's often a dangerous guessing game trying to figure out how to appease them. In many holy books, the *Bible* included, humans are judged to be so flawed, bloody sacrifice and intermediaries are considered necessary to make it possible for God to deal with us.

V. Inside Gods

Depending on your conditioning, you may already be aware of much of what we've said concerning outside gods, but the concept of inside gods may not be as familiar. That is probably because inside gods were an ancient model that was regularly supplanted by belief in outside gods. The terms 'animism' and 'pantheism' both describe inside gods, and they have very similar definitions. Animism claims there is no separation between the spiritual and material world, meaning that everything in existence is permeated by the Divine. Pantheism is a Greek word derived from 'pan' meaning all, and 'theos' meaning god, or quite literally 'all god.' In other words, animism and pantheism are the belief that God and the Universe are one and the same, which is also the basic premise of an inside God.

If an inside god *is* the universe, how was it created? Like all living things on earth, inside gods are said to create, or give birth to creation, out of Self. Instead of 'spirit' being a separate tool or substance used to give and take life, the inside god *is* both the life giving spirit that continually permeates and sustains everything in existence *and* the materials used in construction. That's not at all a strange idea when you consider the fact that a baby is the result of the egg and sperm manufactured by its parents' bodies, and it is nourished in the womb by its mother's body. On a universal level something very similar takes place, energy is transformed into matter and matter into energy, but nothing ever disappears. The same would hold true of an inside god; Source energy can appear as matter, and that matter could return to energy, but it all remains the eternal Divine. This scenario does not allow us to humanize the Divine since Source would simultaneously be everything that exists and the 'nothingness' of all potential. Does this mean that an inside god would serve merely as a ground or matrix that binds and sustains the universe, or could a personality be involved? We'll see what sages and scientists have to say about that shortly, but for the sake of our comparison of inside and outside gods, let's assume that inside gods do have intelligence.

It would be as impossible for an intelligent inside god to be distant, disinterested or uninvolved in creation as it would for intelligent humans to live in detached ignorance of their own bodies. Material creation would have no reason to fear an inside god because harming creation would mean harming Self. And since everything in existence would be an indivisible whole, a part could never be eliminated or harmed without harming the whole. Worshiping, obeying or placating such an inside god would be unnecessary since there would be no need to fear either separation or punishment.

Since an outside god creates from materials that already exist, their creations are not only separate from them; they could be very different from them. This separation and difference allows for the struggle between good and evil without harm to the God who presides over it. However, an inside god, creating out of Self, would be limited to its own 'ingredients.' If you had flour, sugar, butter, salt and water, you could make a variety of different edibles, but it would be impossible to use those ingredients to build a reliable boat or a structurally sound bridge. And we don't expect to harvest oranges from onion seeds or see bears give birth to dolphins because we understand the principle that like comes from like. This natural law means that *everything* that is birthed/created retains the essence of its originator. With an inside God, creator and creation are one, interconnected and inseparable.

An outside god can condemn its creation and label humans imperfect or sinful because they will always be separate and expendable. Since the creation is never quite sure where it stands with an outside creator, it makes the love that is usually demanded difficult to maintain. After all, can an emotion that results from fear actually be considered love? An inside god cannot be separate from creation, and condemning creation would amount to condemning Self. And while an outside god would be impervious to the misery caused by a cosmic struggle between good and evil, this would be impossible for an inside god. Like a rampaging cancer that devours its host, an internal war would ultimately prove to be fatal to an inside god.

This brings up another question: An outside god could easily have an array of emotions that range from love to hate, but could the same be true for an inside god? Since an inside god would be destroyed by an internal war between good and evil, it follows that hateful feelings could be just as self-damaging. It should come as little surprise that sages who have experienced the Divine testify over and over again that "God *IS* love." When they made that statement, they didn't mean that Ultimate

Reality behaved lovingly *sometimes*; they meant the very *Being* of the Divine *is* love. In other words, Source is love and nothing else. Any other positive qualities attributed to Source, such as joy, peace, mercy, compassion or kindness, are all attributes of love. Their words are consistent with the necessary makeup of an inside god, and also tell us such a god displays both intelligence and feeling.

By now the differences we've described between inside and outside gods has probably brought up many questions. The world we see, where good and evil do appear to be continually duking it out, seems to argue in favor of an outside god. And it's true that a majority of religions do believe in the dualistic approach and worship an outside god. However, it's impossible to experience fearless spirituality if God engenders fear. The myriad sages who have lived fearlessly consistently describe the Divine as a constantly compassionate inside god. Which way is it?

There's no doubt that we do see evil and suffering on the earth. In fact, belief in an outside god that is both loving and hateful grew out of the need to give meaning to the struggle we see taking place on the earth. However, as you will see as the book progresses, what we see on earth *does not preclude the existence of an all-loving 'inside' God.* Bear with us, and we'll get to the scientific evidence behind this statement and, more importantly, a spiritual explanation. Indeed, fearless spirituality comes from the understanding that there is *nothing* to fear.

Several paragraphs back we used a quote from Plotinus that referred to Ultimate Reality as "the One." Plotinus used those words specifically to make the point that the Divine is not separate from creation, but permeates it, and moreover IS everything in existence. If you've been taught that God is a powerful supernatural entity that exists outside material creation, this new thought may be difficult to grasp. Is it just a matter of choice whether we believe in an 'inside' or an 'outside' god? Or, is there a way to know if the Divine is inside or outside creation? After examining the difference between the two, it's obvious that our choice cannot help but carry significant ramifications. Of course our view of Source is always our own choice. However, recent research in the field of quantum physics gives us some weighty facts to ponder.

VI. The Quantum Paradigm

Physics is the branch of science that studies energy and matter and their interactions.

Classical or "Newtonian" physics (based on the discoveries of Sir Isaac Newton) studies phenomena such as velocity, momentum, and gravity that explain how the visible portion of the universe operates. Quantum comes from the Latin word *quanta*, that describes the small increments, or parcels, energy can be divided into at the subatomic level. Quantum physics continues to study the interactions of energy and matter, but does so at the subatomic level.

Newton's stable laws allowed scientists and inventors to start thinking in terms of 'reverse engineering.' They imagined that the material universe was constructed of smaller and smaller building blocks, much like a literal building. They began taking apart the material world to ascertain the mechanisms that operated it. Once discovered, they felt certain these mechanisms could be reassembled in new ways to create technologies that could harness and improve on nature. Physicists had assumed that the subatomic world would be governed by the same laws as the visible world, but they were shocked to realize the unseen portion of the universe is remarkably different.

As far back as ancient Greece and India, it was assumed that atoms, from the Greek *atomos*, meaning indivisible, was the smallest building block in nature. However, scientists in the late 1800s began to realize that atoms were also made up of even tinier, subatomic particles. But instead of the separate and distinct subatomic building blocks they expected to find, quantum physicists discovered a sea of interconnected energy where no separation or differentiated form existed. As author Lynne McTaggart pointed out in her book, *The Field*, "The world, at its most basic exist[s] as a complex web of interdependent relationships, forever indivisible." If you were granted special vision and could see the universe at this elemental level, you would find that *everything is one thing!* Subatomic particles, like the cells of the body, have no meaning if isolated from one another. As physicist Fritjof Capra pointed out in *The Tao of Physics*, "Quantum theory...reveals a basic oneness of the universe. It shows that we cannot decompose the world into independently existing smallest units."

Although the oneness found at the quantum level argues in favor of an inside god, you may still feel that this evidence does not preclude the possibility of a god that exists completely outside the material universe. There is, however, another compelling discovery made by quantum physicists that tips the scales in favor of an inside god. Classical physics had conditioned scientists to assume most energy and matter was unconscious. Scientists saw themselves as objective observers who measured and collected the data that resulted from their experiments. They trusted their conclusions because the inanimate matter they worked with guaranteed that successful experiments could be duplicated over and over, by any scientist, with the same results. Imagine their surprise when quantum experiments gave different results depending on which scientists were conducting them!

Quantum physicists had to face the fact that they were no longer objective observers working with inanimate, unconscious matter. Instead, they were active participants in their quantum experiments since even the tiniest particles were alive and consciously reacting to the scientists themselves. How could that be? Quantum particles exist in a state of 'potential.' This means they won't be in a set or stable state *until* they're influenced by consciousness. In other words, the conscious choices of the experimenter are an unavoidable component of quantum experiments. *This could only happen because all energy and all matter is alive and conscious.* Yes, the level of intelligence or awareness may differ significantly, but life and consciousness remain one.

We might assume that scientists were thrilled to discover the entire universe is conscious, but that has seldom been the case. Physicist Peter Russell pointed out in his book *From Science to God: a Physicist's Journey into the Mystery of Consciousness*, that, "Nothing in Western science predicts that any living creature should be conscious." Consciousness is an enigma that science would very often prefer to ignore, especially since its ephemeral nature makes it virtually impossible to measure. When scientists known as 'material realists' claim that only matter exists, they're actually saying that consciousness is no more than another form of evolved matter. Material realists have been unable to discover a shred of evidence to support their postulate (an assumption that's impossible to prove), but the fact that consciousness does exist at the quantum level demonstrates that it has always been present, and didn't evolve. Peter Russell went on to note, "The continued failure of these approaches...suggests [scientists] may be on the wrong track. They are all based on the assumption that consciousness emerges from, or is dependent upon, the physical world of space, time and matter...I now believe...we should be developing a new worldview in which consciousness is a fundamental component of reality." Simply put, matter came from consciousness, not consciousness from matter!

Regardless of the personal beliefs and preferences of some scientists, consciousness carries massive implications for spiritual seekers; implications that we can count on. In their book *The Quantum Enigma*, physics professors Bruce Rosenblum and Fred Kuttner pointed out that whether scientists like it or not, "Quantum theory is not just one of many theories in physics. It is the framework upon which all of physics is ultimately based... Quantum mechanics is stunningly successful. Not a single prediction of the theory has been wrong."

To further the argument in favor of an 'inside god,' physics has discovered that communication between photons (subatomic packets of energy) is instantaneous, taking place faster than the speed of light. Since nothing in the material portion of the universe can travel faster than light, this argues in favor of *one* consciousness that permeates and connects the entire universe. Renowned physicist Erwin Schrodinger agreed saying,

"Consciousness is a singular for which there is no plural." Quantum discoveries such as this have moved many physicists to feel certain that *the universe itself is one, all-encompassing intelligence.* Since we assume we're individuals with private thoughts, how do we fit into this universal 'one mind' model?

It's true that others don't have access to our brain processes, but the concept of hidden and private thoughts is another misperception. If all our thoughts were contained within our brain, our thoughts might well be private, but science is demonstrating they are not. After thirty years of study, neurophysiologist and neurosurgeon Wilder Penfield concluded the brain is a "data bank for information retrieval...there is nothing in the in the brain to account for the high level of experiences and capabilities of the mind." Penfield clearly saw the brain as a control center for the body, but it was not the place where consciousness (the mind) existed. If consciousness isn't part of the brain, where is it? In her book, *Infinite Mind: The Science of Human Vibrations of Consciousness*, researcher Dr. Valerie Hunt answers, "The higher level mind seems to be outside the domain of material reality...the mind is more a field reality, a quantum reality."

Simply put, your brain and the body's senses navigate the material portion of the universe, but your mind/consciousness is within the quantum field that permeates the universe.

There's no doubt that the visible, material portion of the universe *appears* to operate very differently from the quantum world, but physicists tell us that the difference is no more than an appearance. In the Peter Russell book mentioned earlier, Rosenblum and Kuttner point out, "Quantum theory is basic to all of physics, all of science, and is applied to entities as large as the universe and as intimate as the mind." Instead of separate forms that exist in time and space, our bodies are just as much a part of the sea of interconnected energy as a subatomic particle. If our senses allowed us to see the material world as vibration rather than the illusion of solid matter, we would understand. At the subatomic level where scientists thought they would find separate and distinctive building blocks, there is an energy field without boundaries or divisions of any kind. There is no way to tell where one form ends and another begins, and there is no difference in the energy. In other words, human, dog, cat, tree or star energy is exactly the same, and one consciousness permeates it all. Nothing that appears in material form has the power to change that fundamental truth. Since the oneness of everything in existence is our reality, why do we see a material universe of separate forms?

Physicists explain that quantum energy exists in a state of 'potential' known as a 'wavefunction.' When consciousness focuses on this energy, the potential/wavefunction 'collapses' and becomes 'set' as a particle. When that happens, our brain and senses interpret the collapsed potential as material form. This might be easier to understand it we imagine a kitchen stocked with cooking ingredients. Each ingredient has the potential to become part of an appetizer, a main course or a dessert, but they remain in a state of 'potential' until the cook makes a conscious decision, takes an ingredient from the shelf or refrigerator, and mixes it with other ingredients. At that point, the potential of the ingredient ends and it becomes set as something else. Amazingly, before consciousness acts on potential the wavefunction cannot be said to be any particular place! And when consciousness no longer continues to maintain that form, the energy that had been 'set' in particle form once again exists as a wavefunction. This means that the material world isn't actually made up of separate or solid forms, but exists as a 'thought image' that's held in consciousness. When you think the body is feeling, tasting, hearing, seeing or smelling something 'real,' it's actually a sensation (illusion) produced by the brain, not something that exists in reality. This becomes even more mind boggling when you realize *the brain that's interpreting information from the senses is itself a product of consciousness!* Here are a few more thoughts from scientists who came to understand the true nature of life.

Every man's world picture is and always remains a construct of his mind, and cannot be proved to have any other existence...We do not belong to this material world...We are not in it; we are outside. We are only spectators—Erwin Schrodinger

In removing our illusions we remove the substance, for indeed...substance is one of our greatest illusions...Matter is mostly ghostly empty space—Sir Arthur Eddington

I regard consciousness as fundamental. I regard matter as derivative from consciousness—Max Planck

Everything we call real is made of things that cannot be regarded as real—Niels Bohr

The atoms…themselves are not real, they form a world of potentialities or possibilities rather than one thing of facts—Werner Heisenberg

Everything in life is vibration—Albert Einstein

Matter is not made of matter—Hans-Peter Dorr

Although we may be more comfortable maintaining the misperception that we are these material bodies, *physics leaves us with no other conclusion than the fact that we are not who or what we've been conditioned to believe we are.* Instead of being a material body with a personality that thinks with a brain, you, like the universe itself, are pure consciousness that uses energy to create the *appearance* of a material universe. *The real you is not, and has never been, a separate material or supernatural body.* And unlike a body or a spirit being that can only be in one place at a time, pure consciousness is 'non-local.' In other words, consciousness, like energy potential, permeates the universe without a set location. As Peter Russell pointed out, "Our perception of the world has the very convincing appearance of being 'out there' around us, but…all experience is an image of reality created in the mind." (Remember that when the word 'mind' is used in connection with consciousness, it is NOT referring to the brain.)

This is not to say that the material universe is purely illusion, it does have a level of existence within consciousness. However, it is not what it appears to be, and most important, it is *not* our reality. Peter Russell puts it this way, "Everything we experience is a construct within consciousness. Our sense of being a unique self is merely another construct of the mind." In other words, *we are not in space and time, space and time are inside us. We are not the product of space and time; it is the product of our consciousness.* This is not to say that as pure consciousness you have no sense of self, but it is not the self you think of as the body or the personality your consciousness is currently projecting. This might be easier to accept if you think of the material portion of the universe as a fantastic 3D virtual reality that's imagined and controlled by consciousness.

Of course the concept of our world as a self-created virtual reality completely changes our understanding of it. After all, if our conscious thoughts are creating it, we can no longer see ourselves as victims in a struggle between superhuman powers of good and evil. Instead, we're more like players in a complex game or drama that we ourselves have constructed. Later in the book you'll discover why we've done this and how we can free ourselves from this imagined misery, but for now, let's discover what spiritual sages have had to say about this world and our quantum Reality.

VII. Oneness

The model of universal Oneness supported by quantum research is among the oldest spiritual concepts on record. As we mentioned earlier, animism and pantheism were ancient forms of belief in an inside god. Some of the earliest writings that explain universal Oneness are found in a collection of poems and dialogues called the *Upanishads*. Some scholars believe these writings may date back as far as 6000 BCE. Regardless of the exact date, it's clear they were composed by both male and female sages over a period of several hundred years. The writings originated in 'forest academies' along the banks of the upper Ganges River in India. The Sanskrit title loosely translates as "sitting down near," and brings up the image of students sitting at the feet of a teacher. But in this case the purpose was not so much instruction as inspiration, and each participant was expected to learn through their own personal experience.

Since the sages who composed these gems wanted us to concentrate on the message rather than the messenger, they were written anonymously. Although the *Upanishads* are associated with the *Vedas* (the oldest Hindu scriptures) they are free of Vedic ritual and stand on their own as wisdom sayings. As Eknath Easwaran, a 20th century mystic and scholar explained in his translation of the *Upanishads*, "[They] place us at home in a compassionate universe where nothing is 'other' than ourselves—and they urge us to treat that universe with reverence, for there is nothing in the world but God."

Although the *Upanishads* were written thousands of years before quantum research existed, the two complement one another. The *Chandogya Upanishad* tells us, "In the beginning was only Being, one without a second. Out of himself he brought forth the cosmos and entered into everything in it. There is nothing that does not come from him. Of everything he is the inmost Self." Most importantly, the *Upanishads* goes on to tell us, "You are that. . .you are that."

Instead of approaching the Divine with preconceived notions, the *Upanishad* writers displayed a willingness to empty themselves of their misperceptions and be taught by Ultimate Reality. As we look at a few more excerpts from the *Upanishads*, you'll get the gist of what the writers learned when they personally experienced Source.(In the Upanishads, capitalized words such as Being and Self refer to Ultimate Reality.) The first lines are taken from the *Aitareya Upanishad*:

Before the world was created, the Self alone existed;

Nothing whatever stirred.

The Self thought: "Let me create the world."

He brought forth all the worlds out of himself.

As the poem opens, Ultimate Reality exists alone in a state of pure consciousness and potential since "nothing whatever stirred." The key here is the word nothing, which can also be thought of as 'no thing.' No separate material 'things' existed, but the quantum sea of potential, like the ocean, remained a continually seething maelstrom of activity. Of course the next words, "Let me create the world" are simplistic, but they clearly make the point that conscious thought instigates creation.

As we mentioned earlier, even though the *Bible* features an outside god, the creation story found in Genesis also gives a description of the material universe coming into existence through consciousness. In the first chapter of *Genesis*, we find the creative statement, "Let there be..." repeated eight times, and after each statement new material forms came into being. The portion of the *Aitareya Upanishad* you just read goes a step farther than the *Bible* when it implies that not only is the Divine pure consciousness, it is also the energy that consciousness interacts with to bring form into existence. In other words, Source *is* both the consciousness *and* the infinite energy potential that comprises the foundation of the universe.

The *Chandogya Upanishad* adds more detail and tells us exactly what this means for each of us:

In the beginning was only Being,

One without a second.

Out of Himself He brought forth the cosmos

And entered into everything in it.

There is nothing that does not come from him.

Of everything he is the inmost Self.

He is the truth; he is the Self supreme.

You are that...you are that.

The writer explains that everything in existence not only came out of Source, but continues to be permeated and sustained by the Divine. This means that everything, on every level of the universe *is* Source, including you. The conclusion: *you are that—you are Divine!*

If you're used to the concept of an outside god, the thought that you are Divine may sound blasphemous. Like Jesus, the 9th century Sufi mystic, Mansur al-Hallaj, paid with his human life because he openly declared Oneness with the Divine saying, "There is nothing...but God." In the *Bible* some angry bystanders were about to stone Jesus, "...for blasphemy; because you, being a man, make yourself God." (John 10:33) In the gnostic *Gospel of Thomas* Jesus explained oneness this way, "God's kingdom is inside you and outside you. Whoever knows oneself will find this." In the *Bible*, he explained that God's kingdom is not a government, but the Being of God within each one of us when he said, "The kingdom of God is not coming with your careful observation nor will people say, 'Here it is' or 'There it is,' because the kingdom of God is within you." (Luke 17: 20-21)

But Jesus and al-Hallaj were far from the only ones who recognized the Oneness of all that is and our own existence within Divine Oneness. Over the last several years we've collected hundreds of quotes by spiritual sages from all time periods and areas of the world that testify to this Oneness. Here are several that will give you a taste of how pervasive this understanding has actually been among those who have experienced the Divine first-hand. They are presented in order from ancient to modern:

The Tao...is empty yet inexhaustible, it gives birth to infinite worlds—Lao Tzu, Taoist philosopher and author of *Tao Te Ching*, 6th century BCE (The Tao signifies All That Is)

Those who see the consciousness within themselves is the same consciousness within all beings attain eternal peace—*Katha Upanishad*, most likely composed around the 5th century BCE

Emptiness does not mean non-existence. It means that nothing exists independently. Emptiness means empty of separate self—Siddhartha Gautama (Buddha), spiritual teacher, India. Lived sometime between 563-400 BCE

From all things one and from one all things—
Heraclitus, Greek philosopher, 535-475 BCE

What makes things finds no border between
itself and things. If things have borders they
are borders made by words—Chuang Tzu,
Chinese Taoist philosopher, 4th century BCE

He who sees the Lord the same in every creature
sees the deathless in the heart of all who die—
Bhagavad Gita, Hindu scripture written in India
sometime between 200 BCE and 200 CE,
author/s unknown.

God is not external to anyone or anything, but
exists in everyone and is in all things—Plotinus,
pagan Greek metaphysical philosopher and
teacher of 'The One' 205-270 CE

One thing, all things: move among and
intermingle without distinction. To live in this
realization is to be without anxiety—Jianzhi
Sengcan, aka Kanchi aka Sosan or Seng-Ts'an,
3rd Zen Patriarch and author of *Hsin Hsin Ming*,
China, died 606 CE

There is one Reality—like a brimming ocean in
which all appearances are dissolved. How can
it be divided? —Shankara, mystic, teacher and
author of *The Crest Jewel of Discrimination*.
India, 788-820 CE

Every grain of matter, every appearance is one with Eternal and Immutable Reality!—Huang Po Chinese Zen Master, died 850 CE

As to Reality: Reality has no opposite and cannot have, for All is One...one matter, one energy, one Light, one light-mind, endlessly emanating all things.—Jalal ad-Din Muhammad Rumi, among the world's most beloved mystic poets. Persian, Sufi 1207-1273 CE

The day of my spiritual awakening...I saw all things in God and God in all things—Mechtild of Magdeburg, Medieval Christian mystic, nun and author of *The Flowing Light of Divinity*, Saxon 1207-1282 CE

God is unified oneness—one without two, inestimable...constituting a chain linking everything—Moses de Leon (Moshe ben Shem-Tov) Spanish rabbi and Kabbalist 1250-1305 CE

God and I are One—Meister Eckhart, Dominican friar and mystic, Germany 1260-1327 CE He was tried as a heretic by Pope John XXII.

And I saw no difference between God and our substance; but, as it were, all God—Julian of

Norwich, Christian mystic, English 1342-1416 CE (Despite the name, Julian was female.)

All things are living, even stones. It has to be that way...since all are part of the Omnipresent Living Being—Hafiz, mystic poet and Sufi master, Persia 1320-1389 CE

Shame on such a morality that fails to recognize the Eternal Essence that exists in every living thing—Arthur Schopenhauer, German philosopher and mystic 1788-1860 CE

The world is no more than the Beloved's single face—Mirza Ghalib, Turkish mystical poet, 1797-1869 CE

All are one and the One is all—Nisargadatta Maharaj, Indian philosopher, teacher and author of *I Am That* 1897-1981CE

God is not something that transcends reality; God is the base of reality—Nishida Kitaro, Japanese philosopher and professor, 1870-1945 CE

At the center of the universe dwells the Great Spirit, and this center is everywhere, it is within each of us—Black Elk, Native American Medicine/Holy Man of the Oglala Lakota (Sioux) tribe, 1863-1950 CE

Every individual IS the Divine, is infinite, is God; and this moment is eternity—Alan Watts, British philosopher, spiritual teacher and author of numerous books 1915-1973 CE

You are within God and God is within you. You could not be where God is not—Peace Pilgrim (Mildred Norman), American mystic and spiritual teacher who walked across the U.S. for 28 years with her message of peace, love and oneness. 1908-1981 CE

VIII. The Grand Illusion

The Oneness of All That Is has been understood and recorded by mystics and sages for millennia, and their words inspired many who then willingly set out on their own journey of Self-discovery. But more often, humanity has accepted the conditioned responses taught by society, and believed what their limited senses have told them. When we cling to preconceived notions, we can hear the teachings of a sage yet fail to absorb or understand them because the bucket/brain we talked about earlier is still full of stagnant information. At best, all that will result is a muddy mixture that keeps us in a state of confusion. That's why the teachings of spiritual masters often seem garbled and contradictory after they've been muddied and then passed along by those who couldn't grasp what they were hearing.

Nonetheless, it's impossible for our blindness to negate the truth; we can ignore truth, or even fight against it, but we can't change it. In the *Chandogya Upanishad*, a father demonstrated this fact by asking his son to put some salt in a cup of water and let it stand overnight. In the morning, the son was convinced the salt had disappeared until his father asked him to taste the water. As the father explained, the way the salt had dissolved in the water symbolized the way Ultimate Reality permeates the universe; always there, but unseen until we each *choose* to see. To know the Divine, we must taste for ourselves just as the father explained:

It is everywhere, though we see it not,

Just so, dear one, the Self is everywhere,

Within all things, although we see him not,

There is nothing that does not come from him,

Of everything he is the inmost Self.

But the father went on to tell his son that we can be easily tricked by the visible/finite level of the universe and forget that the invisible/infinite realm is our Reality:

Where one realizes the indivisible unity of life,

Sees nothing else, hears nothing else,

Knows nothing else, that is the Infinite.

Where one sees separateness, hears separateness,

Knows separateness, that is the finite.

The infinite is beyond death, but the finite cannot escape death.

IX. The Implicate and Explicate Universe

Physicist David Bohm recognized that even though the universe remains one interconnected, indivisible, whole, it is easier to understand the visible and invisible portions if we describe them as two levels. Bohm called the invisible, subatomic portion of the universe the 'implicate order,' taken from the Latin *implicatus*, to entwine or enfold. This was because Bohm recognized that all potential and consciousness originated within, and permeated, this level. He called the visible portion of the universe the 'explicate order,' from the Latin *explicatus*, meaning unfold, because material form 'unfolds' out of the consciousness and energy potential at the subatomic level. We could think of the implicate order as a seed that contains everything that's needed to create a plant, and the explicate order as the seedling that unfolds from, and is fed by, the nurturing environment of the seed. The seedling is not aware of the seed, yet it could not exist without it, just as we do not see the implicate order even though all life comes from it and depends on it.

What Bohm and many other physicists have to say can be quite startling, especially when they emphasize the fact that the implicate order is our reality, not the explicate order that arises from it. The *Ashtavakra Gita*, written by an anonymous sage around the 8th century, agrees, saying "You are unconditioned, changeless, formless...You are Consciousness...Awareness alone. The world is a passing show...That which has form is not real. Only the formless is permanent. Once this is known, you do not return to illusion." Or as Alan Watts put it, "What you are basically—deep, deep down, far, far in—is simply the fabric and structure of existence itself."

X. The Journey

You got to be careful if you don't know where you're going, because you might not get there— Yogi Berra

As silly as this quote sounds, not knowing where we're going is one of the greatest obstacles to fearless spirituality. 'Journey' is a term that's commonly used in spiritual circles, but our misperceptions about who and what we actually are often lead us on one lengthy detour after another or keep us trapped in the same rut. After all, if you don't know your point of departure or your intended arrival destination, it would be impossible to even plan a trip, let alone successfully reach your goal. Or if we're speaking about spiritual healing, how would we know what the cure is if we can't identify what spiritual health looks like or diagnose the problem we're having? For that matter, we can't 'transcend' if we not certain what we're trying to rise above. Looking back, it became clear to us that not understanding where we were coming from or where we were going sent us on a long series of frustrating detours.

First let's clear up what the word 'journey' means in a spiritual context. The dictionary tells us it's the act of traveling from one place to another or the distance we travel. Either way, the word implies that we are not where we want to be and must take some action to get somewhere else. This view often gives seekers the feeling that they need to be in a 'different place.' Some believe this literally means going on a pilgrimage, traveling from one guru or teacher to the next or transitioning though several spiritual practices. But the second dictionary definition describes a journey as a "process or a passage." This definition helps us understand that a journey can also be figurative; an inner voyage that can be taken without physical action of any kind. Later we'll go into depth on the topic of action vs. process, 'doing' vs. 'being,' but for now, let's address the fact that an internal process, like physical travel, requires a starting place.

You probably take it for granted that the starting place of every journey you embark on is your present location. But in light of your new quantum awareness you do need to ask, "Where am I right now?" Your answer to that question will influence your entire spiritual journey. Unfortunately, religion and philosophy both get fairly muddy when they try to answer that question. If you believe in an outside god, where are you starting? Are you human, part human and part spirit, or a spirit having a human experience? Did you start out in heaven or did your life begin on earth? Which of these places will be your starting point?

If you believe in an outside god, what is your journey? Is it toward heaven or away from hell? Does transcendence mean overcoming a sinful body or trading mortality for immortality? Do you want to have a mystical experience, attain an altered state of awareness, gain more peace, be more loving, come to a better understanding of God, build your faith in a 'savior,' feel that you are 'saved' or find forgiveness? We're asking you these questions because we realized in retrospect that had we carefully questioned ourselves this way years ago, our journey would have been much shorter and far less perplexing than it was.

We spent years desperately trying to know and understand a God we had been taught was outside us. We beat ourselves up with the concepts of sin and guilt and constantly felt that we must be falling short somehow. We learned a great deal about religious beliefs, but felt no closer to the Divine. We failed because *the journey we were trying to make was at odds with what is.* But what we were doing was nothing new; humanity has regularly refused to accept what the universe has been trying to tell us, to our own detriment.

Many of our ancient ancestors did recognize that there was more to the universe than met the eye. They intuitively felt that the gods were an indivisible part of nature. But unfortunately, they began to assign good and evil qualities to the natural phenomenon they experienced. As a result, myths began to emerge that claimed outside gods were creating natural events to either bless or punish humans. As science began to discover a more accurate explanation for natural occurrences, science and religion took issue with one another. Although scientists have explained how and why natural phenomenon take place, many still call these events "acts of God" and blame God for the devastation they can cause. As the gulf between science and religion widened, each side stood its ground instead of investigating what the other had to say. Since Copernicus, Kepler and Galileo (16th -17th century) proved that the church was wrong when it claimed that the earth was the center of the universe, these two opposing factions have been at war.

Although Kepler felt that the universe was created according to a divine plan, his love of truth moved him to write that when it came to nature, truth was more sacred than "the opinions of the saints." Galileo went even further when he wrote that passages from the *Bible* or church dictates should not supersede direct observation of nature or the proven results of experiments. He concluded it was "pure folly" to allow the church to offer opinions about nature. Galileo was tried by the inquisition, accused of heresy, forced to recant and spent the rest of his life under house arrest, but he didn't change his mind. The strength of scientific proof eventually forced the church to admit the earth was not the center of the universe, but the split between science and religion has still not been repaired. The natural world remains the province of science while the spiritual world continues to be the realm of religion. Nonetheless, we all continue to suffer from this needless rift.

Albert Einstein took a more realistic view of the schism when he said, "Science without religion is lame. Religion without science is blind." Einstein realized that when science ignores anything that can't be quantified and religion ignores scientific fact, both fall far short. As physics professor Lawrence Krauss observed, "The universe is the way it is, whether we like it or not." Both religion and science can keep trying to shape it into something that fits their vision, but in the end, it will always remain exactly what it is. The emerging quantum model, proven over and over to be true, presents a compelling picture of the structure and operation of the universe that includes some very tantalizing glimpses of the Divine if we're willing to look. But these discoveries are often startlingly different, and at odds, with the belief systems of most religions. As sincere seekers, we too have to ask ourselves what is more important to us, a cherished belief or truth?

The word religion comes from the French *religare*, and Latin *religio* meaning "to restrain" or "tie back," and the Latin *relegare*, meaning "to read again." These meanings were eventually combined to describe an organized body of believers who agree to bind themselves to a supernatural power, give that power their exclusive devotion, conform to a particular set of doctrines, rules and moral values, and adhere to a specific interpretation of the text that group has designated as holy. Simply put, the starting point of a *religious journey* is agreeing to believe and adhere to a specific set of doctrines that have been compiled by others. *A religious journey begins when you choose which belief system to accept, and then spend the rest of your life conforming to it.* But a *spiritual journey* bears absolutely no resemblance to this description.

If the concept of an inside god resonates with your heart, your journey begins in a very different place. The spiritual masters we've been quoting recommend 'emptiness' as the starting point. They suggest that you 'empty the bucket' of the brain by first letting go of your preconceived notions, attachments and aversions; cleaning house if you will. But what they meant has nothing to do with the religious concept of purifying yourself or seeking forgiveness for sins. And instead of thinking or learning your way to the Divine, these sages tell us the journey is *experiential*. That doesn't mean that we learn through a spiritual practice, but by opening ourselves to the *direct, personal experience of the Divine*. After all, there is no one more qualified to be our teacher. Hafiz, known for his humorous style, put it this way, "Interesting the classroom where God says, 'Forget all that you think you know about Me.' That way some real knowledge might dawn."

And a little more wisdom from a few more of the many sages who recommended 'unlearning' as the first step of the journey:

You must remove your own preconceived ideas of God to actually experience God—Alan Watts

Put your thoughts to sleep...Let go of thinking.—Rumi

Can you step back from your own thoughts and thus understand all things? —Lao Tzu

It is impossible for one who has in his soul any extraneous image to conceive of the One while that image distracts his attention.—Plotinus

God is not found in the soul by adding anything but by process of subtraction.—Meister Eckhart

Forget what you know.—*The Cloud of Unknowing* written by an anonymous 14th century Christian mystic

God, whose love and joy are present everywhere, can't come to visit you unless you aren't there.—Angelus Silesius a Catholic mystic also known as Johann Scheffler, Germany 1624-1677

The 9th century Sufi sage, Bayazid Bistami, explained, "The thing we tell of can never be found by seeking, yet only seekers find it." He realized the concept of a spiritual journey is a paradox. Instead of doing something or going somewhere, it's a process of consciousness, a passage from the brain to the One Mind of the

Divine. And when we've taken this journey, we find ourselves exactly where we've always been. But unless we have the willingness to let go of the brain's incessant conditioning and chatter, we won't recognize where we've always been. If that sounds confusing and contradictory, we promise that it will all become clear and simple as we go along.

Part Two

Misperception's Offspring

I. Letting Go

Madeline owns a bakery, one of the finest in Seattle. Her cakes, cookies and candies are considered artistic as well as culinary masterpieces. Since her creations grace exclusive parties and weddings throughout the city, her reputation is extremely important to her. Madeline has come to the conclusion that her tools have as much impact on the quality of the finished product as the ingredients she uses, so she and her staff treat each utensil with respect. Although she may have paid a great deal for a new pan, cookie cutter or candy mold, if it gets damaged, she has no problem tossing it in the trash. It may not matter to someone making an afternoon snack, but Madeline knows that every treat she makes with a damaged pan, cutter or mold will reproduce the same flaw.

As we learned in Part One, our world view is built on misperceptions that are so ingrained, we rarely notice them. Like Madeline's damaged tools, our misperceptions can only reproduce more misperceptions. This may not always matter, but its importance can't be overstated when it comes to accurately answering life's 'big' questions. As long as we assume that this world is our reality, our attempts to find genuine meaning and purpose will be frustrated. We can keep busy with self-improvement, be of service to others and live a 'good life,' but as enjoyable as those pursuits might be, they cannot help us see beyond the material realm and discover our essential nature. While misperceptions can inspire some intriguing myths, those stories can't accurately explain why we project the world of form. No amount of self-improvement or good works can help us understand why the universe operates as it does. And most important, misperception will never lead to spiritual fearlessness.

The successful spiritual seekers we mentioned in Part One were not satisfied with living a 'good life.' They realized that no matter how well things went for them personally, they couldn't rest until they understood why injustice, suffering and death were inevitable parts of this otherwise glorious world. They wanted a solid foundation that would allow them to 'know' rather than guess. When the second-hand information they had collected failed to answer their spiritual questions in a meaningful way, they intuitively knew there had to be an explanation for the incongruity of our world. Their inner certainty gave them the impetus to toss aside the 'dented pan' of misperception so they could see what was hidden behind their conditioning. If we want to do more than live a 'good life,' we will need to do the same.

Letting go of misperception is a very simple concept, but it can be easy or difficult depending on how attached we are to our present mode of thinking. When we decide to clean out our basement, attic or garage, we often begin by pulling out everything to see exactly what we have. Usually we're surprised by a few items that had been tucked away and forgotten. The same is true of our misperceptions; we can't see exactly what our conditioned beliefs are or how deeply they're entrenched until we start dragging them out and examining them. Like the junk we've accumulated, it's impossible to decide whether our thoughts are valuable or valueless until we take them out of the dark corners and bring them into the light. Many garages, basements and attics remain a jumbled mess for years on end because the owner feels overwhelmed when they think of tackling the clutter. Instead, they continue to live with the chaos and may even leave it for their heirs to clean up. The same is true on a spiritual basis. Sometimes the mental chaos we find ourselves digging through is the result of misperceptions that have been passed from one generation to the next. Yes, there's work involved, and we're each faced with the choice of putting in the effort or continuing to endure our current conditions.

Religion often gives the impression that if we join a church, we're 'hiring' a mediator in the form of a priest or minister to do the spiritual lifting for us. Sounds simple, but unlike hiring someone to clean out your basement, no one else can see what you've got tucked away in your thoughts. They can't do your sorting for you, and they certainly can't remove your conditioning. In fact, they're far more likely to add some conditioning of their own. To break free, it's imperative that *you* sort through and clean out the valueless misperceptions your brain has filed away.

As you'll remember from Part One, we're 'conditioned' by everyone and everything we interact with. The bigger a part someone or something plays in our life, the more likely it is that we'll soak up some of its influence. Since religion has always been deeply intertwined with the power structures of this world, it comes as no surprise that society generally supports the view that religion is the only gateway to God. We're trained from an early age to assume this is correct, and few spiritual seekers have evaded this conditioning. It should come as no surprise that this major misperception is among the most effective at keeping us from knowing Source. As the 20th century philosopher/mystic Carl Jung pointed out, "One of the main functions of formalized religion is to protect people *against* a direct experience of God." [Italics ours] Religions, especially those that have any type of division between the clergy and laity classes, consider themselves to be mediators *between* God and man, rather than a vehicle *to* the direct, personal experience of the Divine, known as *gnosis*.

After experiencing Ultimate Reality directly, the sage Shankara wrote, "A clear vision of Reality may be obtained only through our own eyes when they have been opened by spiritual insight—never through the eyes of some other seer. Study of the scriptures is fruitless as long as Brahman has not been experienced. And when Brahman has been experienced, it is useless to read the scriptures." Brahman is a Hindu word that carries greater depth than the English word God. It describes and distinguishes the one supreme, universal Spirit that is the origin and support of the phenomenal universe, the Divine Ground of all matter, energy, and being, the essence within everything in existence. Based on his own direct experience Rumi wrote, "You are a sea of gnosis hidden in a drop of dew. What are this world's pleasures and joys that you keep grasping at them to make you alive?"

The 2nd century gnostic sage, Monoimus, said, "Look for God by taking yourself as the starting point. Learn who it is within you...you will find God in yourself... Stop searching for God outside yourself. Look for him within. Thus you will find in yourself a way out of yourself" When we experience the Divine directly, we have no need of a minister, priest, guru or teacher. No one else can direct our experience, and as a result, the 'knowing' that comes to us through that experience cannot be controlled or institutionalized. But in our world, religion claims God as its personal property and takes that supposed ownership very seriously. The 20th century sage Sri Nisargadatta Maharaj explains it this way, "The real Guru is he who knows the real, beyond the glamour of appearances. To him your questions about obedience and discipline do not make sense, for in his eyes the person you take yourself to be does not exist, your questions are about a non-existing person. What exists for you does not exist for him. What you take for granted, he denies absolutely. He wants you to see yourself as he sees you. Then you will not need a Guru to obey and follow, for you will obey and follow your own reality." A guru or spiritual master might inspire a spiritual journey, but they can never take it for you.

Religious beliefs are so pervasive within most societies, they've colored the way we feel about Ultimate Reality even if we've never set foot in a church. If you think of Source as a male father figure, see the world in terms of good vs. evil, celebrate religious based holidays or wonder whether or not you'll go to heaven or hell when you die, religion has had at least a subliminal influence on your thinking. If you're among those who are disgusted by religion's track record, your internal image of God probably includes descriptors like judgmental, vengeful, angry, bloodthirsty, immature, authoritarian or schizophrenic. Mark Twain's opinion of the outside god taught by most religions was so low; he said that he would like to "go to heaven for the climate and hell for the company." As long as our brain continues to create a bond between God and religion, we will blame Source for the failings of religion and we will retain the image of God they have created. The cure for this problem is experiencing Ultimate Reality personally. *You don't want others to decide who you are based on negative hearsay, and neither does the Divine.*

How do we experience the Divine? Instead of looking for something outside us, we go within. The 'philosopher king,' Roman Emperor Marcus Aurelius (121-180 CE), made this concept a cornerstone of the spiritual philosophy he lived by. In the book he originally titled *To Myself*, he wrote, "Dig within. Within is the wellspring of Good; and it is always ready to bubble up, if you just dig...Honor the highest within yourself; it is the power on which all things depend, and the light by which all life is guided... Waste no more time talking about great souls and how they should be. Become one yourself." Theologians argue that religion is valid because it's *based on* the direct experience of a spiritual sage. Although that may be the case, we're all aware that a personal experience can't be 'given' to anyone else. You will never play a musical instrument or ride a bicycle if you read about the experience but never participate. Like Alan Watts pointed out, "To 'know' Reality you cannot stand outside it and define it; you must enter into it, be it, and feel it. Religious ideas are like words—of little use, and often misleading, unless you know the concrete realities to which they refer."

No book, no matter how holy it's considered or how long it's been venerated, can substitute for experience. The followers, who thought it would be wise to build a religion *around* a sage, could not have understood the sage's words. If they had, they would have experienced the Divine for themselves instead of turning the sage into an idol. They would have realized they must save themselves, not turn a master into a savior. In the place of experience, they used intellect to try to 'figure out' what the sage meant based on their own limited perception and colored by their own attachments and aversions. As the Sufi mystic Hafiz wrote, "Everything the great teachers have to say amounts to a single hint." A sage's words hint at the experience, but words, no matter how eloquent, can never express it fully and those words can never make it yours.

Spiritual hints are valuable because they demonstrate that the experience is possible for each of us, but the experience itself is still the only way to know. It's rather like seeing another child riding a bike back and forth in front of your house before you've learned how to ride. Even though you might be afraid, the other child's ability to ride makes it clear that when you try, you'll be able to do it too. As the Zen sage and haiku master, Matsuo Basho (1644-1694) advised, "Do not seek to follow the footsteps of the men of old; seek what they sought." Once you've had the experience yourself, you will be able to recognize what is truthful and what is not in so-called holy writings. At that point you can enjoy the hints that you find, but you won't need them.

Since religion, by its very nature, is an organization of followers who conform to a standardized set of ideas, religion would have no reason to exist if each of us experienced the Divine ourselves. Even Albert Einstein recognized, "The religious geniuses of all ages...know no dogma and no God conceived in man's image." But most of those who have come into contact with a sage filtered what they heard through the belief system already locked in their brain. They tried to pour pure water into a bucket already full of stagnant water, and by doing so; they fouled the teachings of the sage. We may think that meeting or listening to someone like Buddha or Jesus would miraculously change us, but if that were the case, it would mean we no longer had free will. *The Kybyalion*, a text based on ancient gnostic teachings, explains, "Where fall the footsteps of the Master, the ears of those ready for his Teaching open wide," however, "The Lips of Wisdom are closed, except to the Ears of Understanding." Simply put, spiritual awareness requires a free will choice. Until we're willing to let go of conditioned information and replace it with truth, no sage, no matter how revered, can enlighten us.

Followers, who were titillated by a sage's words but too fearful to experience the Divine themselves, chose instead to use the words as means to gather followers, power and control for themselves. If they portrayed the sage as a 'supernatural being' they could then infer that it was impossible for a mere human to experience the Divine. With that lie in place, they could emulate the outward appearance of the spiritual master and claim to be a mediator chosen by the master. Regardless of how they rationalized their belief, *it's impossible to recreate a cause by replicating the effect.* Even though many religions teach that we must have a minister, priest or guru or ride to salvation on a sage's coattails, this flawed belief is another serious misperception that keeps us from finding the answers we seek. The Buddha made this very clear when he said, "All the effort must be made by you, Buddhas only point the way... Be a light unto yourself. Be your own confidence. Hold to the truth within yourself, as the only truth...Doubt everything, find your own light."

The Dialogue of the Savior, an early Christian gnostic text, was found at Nag Hammadi in 1949 but may date back to the late first century. It contains conversations that purportedly took place between Jesus and his disciples. Since the Bible treats Jesus as a superhuman savior, we might assume that the savior in the title *Dialogue of the Savior* is Jesus, but we would be mistaken. In this text, Jesus is a completely human teacher. Instead of drawing attention to himself as humanity's savior, he was reported to say, "Enlighten *your* mind. Light the lamp *within you*...Knock on *yourself* as upon a door and walk upon *yourself* as on a straight road...Open the door for *yourself* that *you* may know what is...Whatever *you* will open *for yourself, you* will open." [Italics ours] Everyone who follows this advice becomes their own savior, thus the title, *Dialogue of the Savior*. These words compliment the recommendation Jesus is reported to have given his listeners in the *Bible* at Matthew 7:7, "Keep on asking, and it will be given to you; keep on seeking, and you will find; keep on knocking, and it will be opened to you." Plainly, Jesus did not mean that he, or anyone else, should do the asking, seeking, knocking or the finding for us. If he had, he would have made that clear by telling his followers to sit still, be quiet and wait.

Countless others who have experienced the Divine have shared the same message. Here are a few more of their thoughts:

The Father's word goes out in the All as the fruition of his heart and expression of his will. It supports all and chooses all...Happy is the man who comes to himself and awakens—*The Gospel of Truth*, probably written by the early Christian gnostic Valentinus (100-160 CE)

One who seeks will find; for one who knocks it will be opened—*Gospel of Thomas*, a collection of very early sayings of Jesus found at Nag Hammadi

If you look for the truth outside yourself, it gets farther and farther away—Tung-Shan, aka Dongshan Liangjie, 9th century Zen master

There is nothing outside yourself, look within...Everything you want is there. Knock on the inner door. No other—Rumi

God builds his temple in the heart on the ruins of churches and religions—Ralph Waldo Emerson, poet and champion of the transcendentalist movement, a belief in the inherent goodness of both man and nature, 1803-1882

The knowledge of the heart is in no book and is not to be found in the mouth of any teacher...Who looks outside dreams. Who looks inside awakes—Carl Gustav Jung, the first modern psychiatrist to focus on the spiritual nature of the human psyche, 1875-1961

But the need to experience the Divine for ourselves creates a conundrum. We all understand the paradox of trying to get a job before we have experience, but not being able to gain experience until we've had a job. We will see the Divine in a completely different way once we've had a direct experience, but getting that experience depends in part on letting go of our erroneous thoughts about the Divine that are already clogging our brain. What can be done? Imagine that you were taking a very important car trip. When you reach your destination, you'll be given a priceless inheritance. To get there, you must travel a narrow coastal highway with steep cliffs rising on one side of the road and a sheer drop to the ocean on the other. Suddenly, you come upon a landslide that has sent enough rocks and tree branches down the cliff to keep you from going on. You could move the rubble, but it would take quite a bit of effort. Since there is no detour, you have a choice to make. You can back up until it's possible to turn around and go home, or you can start heaving debris into the ocean until the road is clear enough for your car to pass.

The Hymn of the Pearl, a beautiful gnostic narrative poem urges us to "Awake and rise from your sleep...Remember you are a child of Kings and see the slavery of your life." There is a priceless inheritance waiting for each of us, but our misperceptions are blocking the road. Plotinus, recognized, "It is *impossible* for one who has in his soul any extraneous image to conceive of the One while that image distracts his attention." [Italics ours] Whether we realize it or not, each day we're deciding whether to toss out our conditioning, attachments, aversions and misperceptions or accumulate more. As the contemporary Zen sage Thich Nhat Hanh (1926-) pointed out, "For things to reveal themselves to us, we need to be ready to abandon our views about them."

Yes, we do need to put effort into changing our thoughts, much like the determination that's needed to learn a new skill or give up a harmful habit. *If we want something different, it's a fundamental rule of life that we first have to think something different, and then do something different.* You may be tired of hearing us say this, but we're repeating this concept over and over simply because all change is fearful to the brain, drastic change is terrifying. Until the brain is convinced that change is in its best interests, it will cling to complacency. When we're confronted by the rubble in the road, it's easy to see the effort that's needed and forget the priceless inheritance. We've spent countless lifetimes in misperception, so the pile of rubble may appear to be overwhelming. On the other hand, it's possible to let go of all those lifetimes worth of debris very quickly when we're willing. And every master who has gone before us has promised that our wildest imagination cannot compare with the treasure we'll actually discover when we get past the valueless junk:

All the best that one might believe of God...will come true. And more—Hafiz

God is not only true and good, he is also beautiful. He creates beauty—for the joy of it—Nisargadatta

When after having sought the One...there is nothing higher, nothing more blessed than this... the Supreme is Love itself —Plotinus

God is everything that is good and the goodness that everything possesses is God—Julian of Norwich

God is voluptuous and delicious—Meister Eckhart

I shall give you what no eye has seen, what no ear has heard, what no hand has touched, what has not arisen in the human heart—Jesus, *Gospel of Thomas*

Now that we realize the importance of letting go of our conditioned concepts of God, let's take a closer look at what the universe itself has to say about Source.

II. Who is God?

From a quantum perspective, everything we see in this world is a projection of consciousness. What we see is what we want and/or expect to see. That may sound quite shocking since we often don't like what we see and it's more likely that we feel like the world's victims rather than its creators. Of course you are not personally projecting everything that manifests itself in this world. However, you are playing a part in the group consciousness that does, and the part you play is far more significant than you could ever have imagined. Stick with us as we take a short detour that will explain what we mean.

As you learned earlier, quantum research has discovered the universe is one interconnected, indivisible whole. Sages have been telling us for eons that this whole *is* the Divine, and now a growing number of physicists agree. Counting from the Big Bang (or the Big Bounce as some physicists describe it), astronomers estimate the material universe is 13.7 billion years old, give or take a few billion. Since material form is the result of conscious thought interacting with energy potential, both had to exist before the appearance of material form. In fact, some scientists are beginning to feel that consciousness is the only thing that exists, and all other things such as space-time, matter and even energy are components of it. As more is learned about the quantum world, convincing evidence is piling up that supports the view of the Divine as a foundation or matrix of consciousness. In this scenario, everything else in existence results from, and is a component of, Divine thought. In other words, *what we call 'life' is a thought held by Source*. Zen sage Nishida Kitaro explained the Divine this way, "At the base of consciousness is a transcendent unchanging reality apart from time... God is not something that transcends reality; God is the base of reality."

Because some spiritual philosophies think of the Divine as a universal ground devoid of personality, you may have heard Source referred to as 'The Void.' This term has caused many to fear that knowing Source means losing yourself in eternal, black nothingness. But in this case, 'void' means quite the opposite. Instead of nothingness; the Divine is the infinite repository of all potential, of all that can be. The word 'void' is used simply to express the thought that the potential has not yet been expressed as form. Although physicists have discovered that even something as miniscule as a photon is conscious, that doesn't mean it carries a level of awareness and intelligence that would allow it to be considered a sentient being. For that to happen, the Divine had to create. But that brings up another question: Since the Divine is complete within Self, why bother to create?

If the Divine were merely an impersonal ground or matrix, it would make no sense for beings with a high level of awareness and intellect to be created. But the fact that 'interactive' beings exist, infers that Source interacts as well. Religions that promote outside gods explain that both spirit and material creatures were made to provide the gods with worshipers, who would praise, obey and serve them. Only an outside god, with human-like emotions and fears, could want or need obedience and worship. For an inside god, who is everything that exists, the thought of worship and obedience are ridiculous. After all, your body is a whole, and you wouldn't demand obedience from your stomach lining or worship from your fingernails! From a quantum perspective, the thought of angelic servants who carry out God's will makes no sense. A being of pure creative consciousness has no need for anyone to 'run errands' since a simple thought would bring what was desired into existence.

Several Eastern philosophies and many New Age gurus teach that Source created to gain experience and understanding of Self *through* creation. They claim that it was impossible for the Divine to know Self without having something to experience and something to experience through. But in this scenario, everything that exists including you, is little more than a puppet that's being used as a vehicle of experience. This means that free will and choice do not exist, and we must endure whatever comes to us so that God can 'feel' it and learn through us. If that's true, we have no other option but to conclude that God enjoys suffering, no matter how horrific it might be. Physical and emotional pain, starvation, poverty, devastating losses, natural disasters, abuse, torture, murder, rape, crime, war, hatred, death, etc., etc., etc., must be fascinating and/or enjoyable to a god that would continue to experience these things over and over for thousands of years. It would also be impossible to say that such a god felt any love for the creation it was using so cruelly. It's true that something has gone horribly wrong on earth, but there is good reason to trust that Ultimate Reality has nothing to do with our experiences in the material realm. Continue reading and you will soon understand why.

There are few misperceptions as common as the belief that God has dualistic qualities and can simultaneously be both loving *and* hateful, merciful *and* demanding, forgiving *and* vengeful. Dualistic gods are seen as the creators of both good and evil, and are said to be the ruler of heaven, and if not the warden, the owner of hell. To further confuse us, dualistic gods are described as just, but are also willing to dish out eternal torment for minor infractions and eternal reward for special favors. But the sages who have experienced the One have a very different tale to tell. Although Meister Eckhart was a Christian theologian, he also experienced the Divine and realized that the church's description of a dualistic outside god was incorrect. Here are a few of the thoughts he shared that challenged the church's teachings and resulted in his being tried as a heretic:

A god that could frighten is not a god—but an insidious idol and weapon in the hands of the insane.

I find nothing more destructive to the well-being of life than to support a god that makes you feel unworthy and in debt.

A god who talks of sin is worshipped by the infirm. I find nothing more offensive than a god who could condemn human instincts.

It is a lie—any talk of God that does not comfort you. How long will grown men and women...keep drawing in their coloring books an image of God that makes them sad.

The Sufi poet Hafiz agreed, saying, "It is a great injustice and a monumental act of cruelty for any religion to make someone fear God." And the Christian mystic, Catherine of Siena, who like Meister Eckhart experienced a very different God than the one taught by her religion, was also brave enough to say, "Strange that so much suffering is caused because of the misunderstanding of God's true nature."

Duality, which inevitably sets up a conflict between opposites, would act like a cancer within the Oneness of All That Is. Just as we would not purposely allow a disease to devastate our body, it would be ridiculous to think that Source would allow the cancer of duality to spread within and inevitably destroy everything in existence. Meister Eckhart recognized this and encouraged us to eliminate the concept of a dualistic God when he said, "Separate yourself from all twoness. Be one in one, one with one,

one from one." If Ultimate Reality is not an impersonal ground, a puppet master or a dualistic schizophrenic, what is the true nature of the Divine? In Part One we briefly stated that spiritual masters have always agreed that "God is love," now let's take a closer look at what this actually means for you and I.

Of course the point of this book is that you leave books behind and experience Divine love for yourself. And as you've learned, you will be open to that love as you let go of dualistic conditioning that claims love can also be hate. The experiences of others can't substitute for your own, but their words can tempt you with a glimpse of what Divine love offers. The ancient *Nasadiya Sukta* or "Hymn of Creation," is part of the Hindu *Rig-Veda* (1700-1100 BCE). It may be one of the first texts to address what took place before the Big Bang. Along with the *Upanishads*, it may also be the first to discuss 'monism,' the view that the universe is One, and everything exists as part of that One. The following are a few of the remarkable lines taken from *Nasadiya Sukta*:

At first was neither Being nor Nonbeing...there was no death...The One breathed without breath by its own impulse; other than that there was nothing at all...Then that which was hidden by Void, that One, emerging, stirring, through power of Ardor, came to be. Desire entered the One...It was the earliest seed, the product of thought...In the beginning Love arose, which was the primal germ cell of mind.

Simply put, the consciousness at the foundation of all that exists, the One, *chose to BE love*. As you'll see, *being love* has no connection with the dualistic transaction of giving and receiving that masquerades as love in our virtual reality.

The first century gnostic Valentinus also described love as the very being of the Divine, and the prime motivation behind creation saying, "Since the Father was creative, it seemed good to him to create and produce what was most beautiful and most perfect to himself. For he was all love and love is not love if there is nothing to be loved." Notice that Valentinus said that Source is *all* love. He wasn't talking about love as an emotion that can be given and withheld or a feeling that could turn to hate. Valentinus meant that love was the Divine's complete and unconditional state of being. Any other qualities demonstrated by Source, such as compassion, joy or peace, are reflections of that love. The mystic poet, Rabindranath Tagore (1861-1941), proclaimed, "Love is the only reality, and it is not a mere sentiment. It is the ultimate truth that lies at the heart of creation." Rumi also recognized love as the being and motivation of Source when he said, "Without Love, nothing in the world would have life." Every sage who has experienced the Divine has reported that "God is love," with the emphasis on *is*.

Why would our Source choose to *be* love? Even though the emotion that pretends to be love in this world is a far cry from the unconditional, unlimited love that Source is, few of us would deny that love is the most rewarding and joyful aspect of life. Source needs nothing from us, but that does not preclude the enjoyment of loving relationships with other sentient beings. As Julian of Norwich put it, "Everything has being because of God's love." The words "God is love" imply that not only does the Divine give love; there must be something to receive that love. Although the Divine doesn't have a *need* to be loved in return, sharing love expands Source's joy. Keep in mind that joy doesn't have a limit, so the more love that is exchanged; the more joy can be felt. As a being that is *only* love, the Divine obviously could never be satisfied by kowtowing robotic beings that had no choice except to love, worship and obey. For there to be meaningful interaction and loving companionship with the Divine, other conscious beings would also have to have a sense of self-awareness, a high level intelligence, and a range of emotion. Since a loving relationship cannot be said to exist without choice, free will would be an essential ingredient.

Many religions tell us that the gods gave both the spirits and the humans they created free will, but also demanded it be used in a specific way or there would be a severe punishment. In the *Bible* book of Genesis, God told the first humans that they could not eat from a specific tree. If they disobeyed, God proclaimed, "in the day you eat of it you shall die." (Genesis 2: 17) In fact, the *Bible* claims that all of humanity was condemned because Adam and Eve used their 'free will' in the 'wrong' way. (Genesis 3: 14-24) But let's face the facts, the moment there are conditions attached to free will; it can no longer be considered free will. That sort of back-handed 'gift' could not help but turn creation into fearful slaves. Free will means that there are no demands or expectations; free will has no conditions or strings attached.

III. The Nature of Free Will

Since free will can have no strings attached, this creates a conundrum. To give the gift of free will meant the Divine also had to allow for, and more importantly *accept*, direct opposition. When we understand that *anything* done within the whole affects *everything* within the whole, free will sounds impossible. Let's begin our discussion of free will by acknowledging that absolute freedom does not exist. This means that all free will is relative, even the free will expressed by Source. That may sound odd, but even the universe must operate within certain operating parameters. Since Source is everything that exists, Source exists within those operating parameters as well. Within these parameters, we're free to make any choice we want, but the choices that defy natural limits would carry consequences. For example, free will allows you to jump off a cliff. You can also choose to fly when you jump, but unless you're attached to some sort of flight mechanism or a parachute, gravity will prevail. Of course it would be great to be able to fly unaided, but we all understand that this desire doesn't fit within the natural laws or parameters that govern the material world. Life in this world would be impossible if these operating parameters, known as the laws of physics, were constantly shifting.

Nonetheless, our free will gives us nearly infinite choice within the parameters that govern and protect All That Is. This is similar to the operating parameters of a car. When you buy a car, it comes with an owner's manual that lists the 'rules' you need to follow to keep your vehicle operating. Still, you're free to make thousands of modifications and your car will continue to function. But no matter how strong your desire, you can't take the engine out of your car or remove the transmission and expect to drive down the street without them. When we speak about free will, we mean choices that fall within the range of the operating parameters that allow Ultimate Reality to continue functioning. (Please keep in mind that these parameters have nothing to do with the moral or religious laws prevalent within religion.) Obviously, if Source could not exist outside universal operating parameters, neither could we.

On the other hand, many religions do claim that creation thwarted the will of the Creator and has been operating successfully outside God's laws for eons. They explain that spirit creatures were able to challenge God's sovereignty, turning themselves into opposers (Satan, devil and demons). These opposers have supposedly been fighting God and recruiting humans to their cause. This scenario is laughable since the rebel creation would not only have to be stronger than the creator, it would also have to have to power to sustain its own life apart from the life-giver. Nonetheless, religion continues to use this impossible story of rebellion and resulting alienation from God to explain the misery that fills the earth. In light of quantum Oneness, this myth is impossible to accept. Still, we're left with two riddles. For free will to be free, it would still have to allow for choices that cannot exist in oneness. How can that be? Second, we do see a world that's out of harmony with the Divine. How can that be?

Once again, quantum physics offers the answer. As you learned in Part One, the universe exists as an interconnected, indivisible whole. You also learned that it appears to have two natures, the implicate/quantum level and explicate/material level. Physicist David Bohm likened the material portion of the universe to a 3D holographic image that's projected into space. A holographic image is an illusion; it appears to be real, but it isn't. On the other hand, the holographic film the image is projected from is like the quantum world. Exposed holographic film looks nothing like photographic film. If you're so far into the digital age you don't know what exposed photographic film looks like, it contains the complete image of a photo. If you cut a photographic film negative in half, you would only be able to process half of the picture from each half of the negative. Holographic film is very different; you won't see the photo replicated on the film. Instead, the image is spread all over the film. If you looked at exposed holographic film, you would see an 'interference pattern' that looks like the intersecting rings that appear on a pond in a light rain. Because the information needed to project the 3D image is smeared all over the film, you could cut the film in many pieces and still project the entire image from any of the pieces. Bohm likened the quantum level of the universe to holographic film for two reasons. First, information on the quantum level has no exact location, but is spread throughout just as the image is spread throughout the film. Second, the hologram appears real but isn't, and the film appears unreal but is. The same can be said of our universe. The material portion appears to be real but isn't, while the quantum level appears to be unreal but, in fact, *is* Reality.

Projected like a hologram in the material realm, the body appears to be an isolated object that can only be in one location at a time. But remember, your reality is not the body. Instead, your reality is consciousness that permeates the quantum level. Consciousness is 'non-local,' meaning it doesn't reside at a specific address. Since your Reality is in the quantum, you are infinite and immortal. Despite what appears to be happening in the material world, your true Self *cannot* come to any harm. However, you do have the ability to project images that, like the 3D hologram, appear to be very real. So real, that we've forgotten who we actually are and believe we are the bodies that are embroiled in every imaginable drama that we ourselves are projecting.

Bohm chose to call the material world a 'holomovement.' Why? Like a movie, each image follows the next so closely; our world appears to be in motion. Many scientists currently agree there is no movement in our 3D virtual reality, but it's actually made up of a series of static images that each resembles a single movie frame. In a sense, it's as if we're playing an extremely sophisticated video game. You choose an avatar to represent you, and as you sit safely on your couch, the avatar has both exciting and horrific experiences, successes and defeats. You feel the excitement as if you and the avatar were one, but the avatar is the one that triumphs or experiences the virtual wound or death. This material 'playground' offers a perfect place to express free will without ever harming what is Real.

The fact that the material world is not our reality confronts us with an extremely powerful misperception. Even though few understand their quantum reality, they do believe that they are a spirit being that's having a human experience. But why would that experience be taking place? A common explanation is that we aren't ready to live as a spirit. We're told that we have things to learn or we need time to evolve, and the material realm is the 'earth school' where we receive lessons and are tested. But we must ask ourselves if it makes sense that a loving God would create us 'half-baked' and then use suffering as a teaching tool? As Alan Watts explained, "The real you is not a puppet life pushes around." Quantum physics makes it quite clear: if we believe we're in this world, it's because our own consciousness is projecting this experience. We weren't sent to earth to learn and we aren't being tested. *Source is not projecting the material world we experience, we are. Why? Because we wanted to express our free will and test Ultimate Reality!*

But what are we trying to accomplish? As you've already learned, Ultimate Reality, and therefore everything in existence, is an interconnected, indivisible whole. There is infinite possibility within that whole, but there are two things that Oneness cannot sustain: separation and specialness. Within Oneness each being can be unique and express their talents and interests in countless ways, but all expressions contribute to the highest good of the whole. Each facet of a diamond contains its own individual sparkle, but all the facets sparkling together make the whole. If you tried to chisel a few facets out of a diamond so they could be separate, the gem's 'wholeness' would be destroyed and the separate facets would lose their value. Oneness has to be that way because the Divine cannot fracture Self into separate pieces, prefer one piece of Self over another, or treat one piece well to the detriment of another. Inequality would set up a conflict that would inevitably destroy All That Is. This is evident even within ecosystems on earth; when one part of the system gets out of balance, it inevitably affects the whole. If things get too far out of balance, the system reaches a tipping point from which recovery is no longer possible.

Imagine separation and specialness as the facets of a diamond each going their separate way, each facet trying to outshine the others! It's an impossible and ridiculous image, but that is what we've tried to do. The hallmark of Oneness is mutual cooperation aimed at the highest good for all concerned, but separation and specialness require inequality, duality, a belief in scarcity and competition. Regardless of the fact that there is no place for separation and specialness within Oneness, we still insisted on giving it a try, and free will provided a safe way for us to experiment. Unfortunately, we've forgotten that this is just an experiment.

When we observe the overall functioning of the entire material universe, it's obvious there's an innate balance that maintains harmony. Astronomer, mathematician and atheist, Sir Fred Hoyle was so astounded by his research he had to admit, "A common sense interpretation of the facts suggests that a superintellect has monkeyed with physics, as well as with chemistry and biology, and that there are no blind forces worth speaking about in nature. The numbers one calculates from the facts seem to me so overwhelming as to put this conclusion almost beyond question." In contrast, the separation and specialness we project on earth has created extreme imbalances and disharmony. Why? Separation and specialness operate from a dualistic system much like the lottery; the majority must lose so a few can win. Specialness can only exist in the presence of 'less than special.' Health can only exist in the presence of illness; wealth requires poverty and happiness demands suffering, etc., etc. Duality is a doomed system; the very nature of its disharmony and imbalances will ultimately bring it to a tipping point and cause it to implode. Still, free will allows us to prove this for ourselves because nothing real is at stake.

An early gnostic text titled *Creation of the World and the Alien Man* pointed out that life in this world is "a dualistic scheme." But it's not the mere existence of opposites that creates the problem of duality; Oneness holds within itself a perfect balance of opposites. Instead, duality is an unhealthy way of looking at opposites. Instead of seeing all opposites as an integral part of the whole, duality sets up a continuum with opposites on each end. The problem begins with the labels we assign to everything we see and manifests when one end of the continuum is desired and its opposite is rejected. Simply put, nearly everyone desires and strives for what we have labeled wealth, power, health, beauty, talent, etc. but rejects what we label as poverty, weakness, ugliness and lack of skill. This results in the desired objectives being scarce and everyone competing for them.

When everyone wants something that's in limited supply, only a few can win while many lose. All areas of life become a quest to see who or what is 'best,' and 'first place' is the only place that matters. Instead of cooperating for the good of all, we feel that we have no choice but to view everyone else as our competition. Love tells us to give, but duality tells us the only way to have what we desire is to take from others. Once specialness is attained, we cannot rest since specialness must be continually guarded and/or upgraded. Mahatma Gandhi, the 20th century Indian leader that modeled non-violence, accurately pointed out that there's enough in the world "for everyone's needs, but not for everyone's greed." It's the belief in separation and the push for specialness that keeps the world in a condition of extreme imbalance.

Is it any wonder that the desire for separation and specialness has resulted in approximately

14,500 wars and only 292 years of peace since 3600 BCE? The wars that have grown out of our intense competition have taken around 4,126,000,000 lives, close to 2/3 of the 2010 world population. It would be wonderful to say that history has made us wiser, but the 20th century and the beginning of the 21st has been the bloodiest in history, with civilians making up well over 60% of war casualties. How did our thoughts go from loving cooperation to lethal competition? What process took us from conscious Reality to virtual reality?

We've learned that our greatest misperceptions include our belief that we are these bodies and the material world is our reality. It's impossible to say how long we existed as pure creative consciousness, but at some point, we developed a desire to trade Oneness for separation. Instead of being unique but equal, we wanted to be special, we wanted to do things our own way. Like a teenager who goes through a rebellious stage, we wanted to push the limits and see what would happen. And, like a loving parent that knows when a young person will only learn by experience, Source gave us the opportunity to try. Since separation and specialness can only exist within a dualistic system, we had to project one where our experiment would be possible. But we didn't stop there; we also wanted to prove that a dualistic system could work as well, or better, than Oneness. Most religions would define those desires as a rebellion worthy of earning us God's eternal condemnation. But when we consider that our challenge is being carried out in a completely harmless way, with Divine support, it becomes little more than a mistake in judgment that can be corrected.

In fact, our challenge was carried out in total trust. Our 'rebellion' was carried out with an escape route and the knowledge that literal separation from the Divine was impossible. Rumi confirmed that our confidence was justified when he said, "The second you stepped into this world of existence a ladder was placed before you to help you escape." Although most religions teach that expressing free will is very risky business, we understood that our challenge was made in perfect safety. We were aware from the beginning that we could stop our experimental 'game' anytime we wanted. Unfortunately, we became so embroiled in the drama we were projecting; we forgot that extremely important fact.

Of course every experiment calls for a laboratory of some sort and research equipment; ours was no exception. It's impossible to say whether the material portion of the universe came into existence specifically for our experiment or if it already existed within our consciousness and appeared to be a perfect environment for our test. Regardless, we needed a 'stage' to play out our drama and an image that we could use like a video game avatar. For the experiment to work, our consciousness also had to be altered. It would be impossible to be completely tuned into the One Mind while also feeling separate from it. To facilitate the experience of separation, it's as if we shut ourselves into a small corner of our infinite consciousness, a corner that imagined it was no more than a body and personality. To differentiate between these two states of consciousness, sages refer to the full consciousness we share with Ultimate Reality as the Self or true Self. The small portion that projects our experiment in duality is known as the self, little self or false self. The little self is never fully disconnected from Self, but it chooses to be oblivious to the connection. Although many teachers use the word ego when speaking about the little self, we prefer not to confuse the two. The ego is the body's own sense of self, something that it needs to successfully navigate this world. Yes, the ego can get into trouble, but without it, the body would find it impossible to respond to the challenges of the surrounding physical and social world. However, the ego associated with each body ceases to exist with the death of that body, while the little self goes on and projects another body.

Since the little self remains purposely ignorant of the true Self, many sages have likened this chosen oblivion to a dream, a nightmare or even a drunken stupor. This comparison makes sense because a body that's asleep or drunk seems to be operating on a different plane than the same body when it's awake or sober. When we're asleep and dreaming, it's not the body that feels real but the dream, just as our projected virtual reality appears to be real. In the gnostic *Gospel of Truth* Jesus observed, "Thus they were ignorant of the Father, he being the one whom they did not see...there were many illusions at work...and there were empty fictions, as if they were sunk in sleep and found themselves in disturbing dreams." In the *Gospel of Thomas* Jesus said, "I found them all drunk...they are blind in their hearts and do not have sight; for empty they came into this world...but for the moment they are drunk." Two thousand years ago, the gnostic author of *Creation of the World and the Alien Man* also confirmed, "Life is a dualistic scheme...a sleep, a drunkenness, an oblivion...our ignorance is a form of unconsciousness."

Since sages view this world as a sleeping or drunken state compared to our natural state of full awareness, it follows that realizing who and what we actually are would be considered 'waking up.' The word 'enlightenment' also attempts to capture the image of 'waking up' by inferring that we are no longer in the dark (sleep) but have come into the light (wakefulness). So, enlightenment can mean both an awakening to the truth that this world is not our reality, and a literal reawakening from partial to full consciousness. In Sanskrit, the word *bodhi* means both 'awakening' and 'understanding.' *Prajna* and *Kensho*, both Buddhist terms, refer to 'seeing/realizing your essential nature or essence.' The ancient *Upanishads* are very clear concerning the difference between our dream state and waking from it. Here are a few samples:

Wake up from this dream of separateness— Shvetashvatara Upanishad

When the wise realize the Self, formless in the midst of forms...omnipresent and supreme, they go beyond sorrow. Those who see the consciousness within themselves is the same consciousness within all beings attain eternal peace—Katha Upanishad

Without the Self, there is no life.—Isha Upanishad

He who finds the Self is free; he has found himself, he has solved the great riddle; his heart is forever at peace. Whole, he enters into the Whole. His personal Self returns to its radiant, intimate, deathless Source—Mundaka Upanishad

Instead of a savior, what we actually need is an alarm clock to wake us from this dream! As Plotinus recommended, "Shake off the fantasies of a dream and bring to an awakened state the awareness which is creating them." Hafiz encouraged, "Awake, my dear. Be kind to your sleeping heart. Take it out into the vast fields of Light and let it breathe." And Rumi added, "One day you will look back and laugh at yourself. You'll say, 'I can't believe I was so asleep.'" But once you do begin to wake up, he warned, "Don't go back to sleep."

As we look at the state of the world we've projected, it's apparent that our experiment is failing miserably. A dualistic system could never rival oneness, and was doomed from its inception to produce exactly what we're seeing.

It's like the dented pan, flawed from the beginning and only capable of reproducing the same flaw over and over again. We've tried every possible combination of political, social, economic and religious systems, but instead of successfully challenging Oneness, they've led us to the brink of destruction. And the odds have demonstrated that duality will always produce far more misery than it does specialness. It should come as little surprise that most of the world's wealth and power is held by a tiny percentage of the population. Since that's the case, why haven't we just stopped the experiment? Since free will is involved, each of us who decided to play this game also has to decide individually that we're ready to stop. We're given the impression that only a very few, like Buddha or Jesus, were enlightened, but when we do some digging, we discover that thousands have already made the choice to return to Oneness. Until we get to the point where we personally decide this world is a failed experiment, we'll keep dreaming and keep trying to make duality work.

No matter what we choose, it's imperative that we remember two things: *everything* we've done has been carried out with the full permission of Ultimate Reality and no real harm has taken place. However, permission is not the same as participation. The point is that we do this on our own. In fact, Divine involvement or intervention would nullify our experiment by interfering with free will. After experiencing the Divine first-hand, Rumi explained, "God is so infinitely tender-hearted and so overflowing with grace, that if He could die for you so that duality could vanish, He would." Unfortunately, free will demands that the only way out of our experiment is through our own choice. Regardless of the truth of the matter, people constantly give God both the credit and the blame for things that happen on the earth. They feel certain that God participates by giving us rules and then blessing or condemning us depending on our level of obedience. But this is no more than a convenient excuse for ignoring our own responsibility. It allows us to take on the role of innocent victim or blessed beneficiary, but in the end, physics demonstrates that our own consciousness is projecting what we see. However, the fact that we are responsible is good news! If our thoughts of separation and specialness can project this world, our thoughts of love and Oneness can also return us to Reality. During the process of returning to the One Mind, we can attain and enjoy fearless spirituality.

Remember that Rumi promised, "The second you stepped into this world of existence a ladder was placed before you to help you escape." When we began our experiment, we needed and wanted an escape hatch. We agreed to the escape 'ladder' being put in place or it would not be waiting for us. We can count on this escape route because it's been successfully used by everyone who has rejected separation. What exactly is it? It's not a secret or anything magical. Since this experiment takes place in consciousness, the ladder into and out of the material world has to be free will choice. Choice caused our dream, and choice will wake us up, no matter how deeply we sleep. In our quantum universe, choice is our most powerful tool. Hafiz put it perfectly when he said, "You carry all the ingredients to turn your life into a nightmare. Don't mix them! You carry all the ingredients to turn your existence into pure joy. Mix them, mix them!" Notice that he did not give the responsibility to anyone or anything outside you, including Source.

To assure free will, Source doesn't interfere by assisting us up the ladder until we've made the choice to set our feet firmly on the first rung. That happens when we're not only willing to see what the Divine has to show us, but moreover, we're willing to accept it. Currently there's a popular movement toward oneness. While oneness is always a positive thought, this movement is focused on what we can do to cooperate with one another while remaining within our dream and is aimed at improving the dream. The ladder Rumi was talking about is there to take us out of the dream/explicate level and returns us to Self/implicate level. When we're willing, our choices will begin to steer us away from illusion and back towards Reality. Those choices are the invitation the Divine and the true Self wait for. From that point on, you can expect to receive the assistance you need to make the climb back to Reality. However, that assistance may not be what you were expecting. Most of us have been trained to ask God for material things, for an improvement in our life situation, physical healing, forgiveness or protection, so it should come as little surprise that many current spiritual teachers also claim that enlightenment means a protected, comfortable life. But what the Divine is interested in is our return, and the assistance we receive will facilitate that return.

Adyashanti, an American Buddhist currently teaching in San Francisco, offered an accurate description of enlightenment when he wrote, "Make no mistake about it—enlightenment is a destructive process. It has nothing to do with becoming better or being happier. Enlightenment is the crumbling away of untruth. It is seeing through the façade of pretense. It's the complete eradication of everything we imagined to be true." As Buddha explained, "I gained nothing at all from supreme enlightenment, and for that very reason it is called supreme enlightenment." Instead of gaining something new, we let go of the valueless so we can see the treasure we have always had. The rewards of waking up are very different than what the brain and body would imagine and desire. *What Self and Source support is your return to Oneness, not improving your experience of separation and specialness*

As we wake up we change inwardly, but that inward change will be so profound, it cannot help but manifest in the material world. That should come as little surprise since we're letting go of the dualistic thinking of the little self and connecting more fully with the true Self/One Mind. Spiritual mastery is not mastery over others, but mastery of the little self. As you live more as Self, the thoughts, desires and actions of the body change. And instead of the body running the show, the Self will use the body as a tool. The body and personality will eventually feel like little more than a helpful garment or disguise that you use to navigate the world. Waking up does not take us out of the world, but we are no longer a part of the thinking that makes it. The chaos of this world will continue to swirl around you, but as the awakened Self, it no longer sucks you into its insanity, nor can it disturb your inner peace or joy. The Divine can and will intervene on your behalf, but that intervention will be focused on bringing you what you need to support your return to Oneness, not keep you here. Once again, Source wants to free you from the prison of duality, not make it more comfortable so you'll want to stay. We've repeated that thought several times, because you can't successfully complete a journey unless you have a specific destination in mind and are determined to go there. Yes, you can continue wandering in this world without a destination in mind, but be aware that this world holds no satisfying destination for you discover by chance.

Waking up is not an end in itself, but the beginning of a process. The ultimate goal is a literal return to Oneness at the quantum level finally ending the cycle of birth and death that keeps us in this dream. Sages have always pointed to this return as our ultimate goal. Rumi advised, "Untie the knot of existence...so you can at last escape the tyrannical false self... I know this world; it never fulfills its promises. What an absurd fate to give your life to this world! Return at last to the origin of your own origin...escape that prison with its thousand traps. One of the marvels of the world is the sight of a soul sitting in prison with the key in its hand." Since we're attached to the idea that we are the body, the personality and its life story, the idea that none of it is real may seem frightening at first. But as we begin to become reacquainted with the Self, that fear melts away. As Hafiz so rightly said, "Listen: this world is the lunatic's sphere...my address is somewhere else."

IV. Transcending the Experiment

Like most people, we thought that security meant having physical protection and material blessings from God. We believed spiritual healing, awareness and transcendence were about self-improvement. In other words, we felt sure we would be 'approved' if we could get our brains and bodies to obey rules we were taught God had written. We were convinced by religion that belief, worship, obedience and sacrifice were all necessary for salvation. And, like most other people, we thought that this was difficult because our natural inclination was toward sin. It took years of struggle before we got to the place where we could see that it was our conditioned misperceptions that were standing in the way, *not* a judgmental God or a sinful nature. Then we understood that misperceptions were possible only when a portion of our consciousness was out of sync with Reality. This realization was shocking, but it also resonated in the deepest part of our being. We had never been able to reconcile the concept of a loving God with the misery we saw in the world, and we were correct to feel that way. Understanding our quantum Reality and the role consciousness plays in projecting the material universe explained why the world is the way it is. Since this change in thinking can be quite overwhelming when you're first exposed to it, let's recap what we've discussed so far before going on to Part Three:

Shankara, like all other masters, realized through his own experience what it takes to wake up. He explained, "Who can be called wise? He who can discriminate between the real and the unreal. What is unreal? That which disappears when knowledge awakes." The fundamental key to enlightenment is *unlearning*, letting go of the misperceptions that keep taking us in the wrong direction. Rumi observed, "Every time you really look at your false self, you die. After thousands of repeated deaths, you begin to realize what in you lives forever." *Our most fundamental misperceptions are hidden in our own concepts of who and what we actually are. We can't see our true identity because we assume this world is real.*

Author Joseph Chilton Pearce got to the crux of the matter when he observed, "That we are shaped by the culture we create makes it difficult to see that culture is what we must transcend." The definition of culture offered by author Terrance McKenna, hits far closer to the mark than most people would care to admit, "Culture, which we put on like an overcoat, is the collectivized consensus about what sort of neurotic behaviors are acceptable." From the time we're born, we're conditioned by everything around us to accept a premise that isn't true. Instead, we've collected an assortment of misperceptions without doubting their credibility. And it turn, we've carried on the misperceptions by conditioning others. The brain supports conditioning by filtering out new information it finds uncomfortable so it can maintain the status quo and find safety in the crowd. Before the brain can make significant changes in its information files, it needs compelling reason to let go of cherished, yet poisonous, misperceptions and the collective mind. This may appear to be both frightening and exceedingly difficult, but countless men and women throughout history have successfully made the choice to transcend the world's conditioning. They have discovered their Reality entwined in the Oneness of All That Is. These seekers were exactly like you. If they could transcend this world, you can too.

What seekers have known through spiritual experience, quantum physics is now confirming. Science and religion have been bitter enemies, but study of the natural world and knowing the Divine are two sides of the same coin. Instead of a God that exists outside the universe, we discovered the universe and Ultimate Reality are synonymous. Instead of being a mortal, material body, we've discovered that we are an infinite, eternal and indivisible part of the quantum consciousness that permeates everything in existence. The material world combines energy with consciousness to project an extraordinary 3D virtual reality. What we see in the world is the result of the thoughts and desires that we've projected. The body and personality are no more than actors/avatars that we use to play out the drama we continually imagine.

Instead of being tested by God, we're the ones doing the testing. We've set up an experiment to prove separation and specialness could rival Oneness, but we've become so enmeshed in our own game, we've lost track of who and what we really are. Our dualistic system was doomed to failure from the beginning, but Source gave us the gift of free will and would not deny us its full use. However, nothing we can project, including the most horrific atrocities, has ever harmed Reality, and you are Real. We have always been safe and always will be, but as long as we remain asleep, we'll continue to believe we're suffering. Choice is involved in every step of the spiritual path. Choice got us where we are, only choice can get us out.

Part Three

The Real Self

I. The 'little self'

Researchers estimate that approximately 60,000-75,000 thoughts make their way through the average brain during a 24-hour period. If we were fully aware of each thought, we would undoubtedly be overwhelmed and unable to function. We've already discussed the fact that the brain 'protects' us by going through a large portion of its sorting and filing work without our direct attention. This is possible since many of our thoughts are slight variations on a few themes that are continually repeated. And although we may believe we're absorbed in meaningful thought, researchers believe approximately 93% of our thoughts are focused on our own personal concerns, needs and desires. For the body this makes sense because the brain's primary function is to keep it safe from all real and perceived danger. As a result, a large percentage of the thoughts that pass through the brain are fueled by varying degrees of fear. The little self uses our non-stop thoughts, and especially our fears, to drown out the truth that can set us free from illusion. But it may surprise you to know that the little self also suffers from its own set of fears.

When we began our experiment, we agreed to an escape route, (remember Rumi's ladder) and knew that we could stop any time we wanted. But somewhere in time, that knowledge was forgotten. Instead of remembering that we were free to experiment, the little self began to believe that this world was the result of its successful rebellion against Source, a revolt the little self reasoned was certain to end in horrible punishment. We all know that the brain can take a small fear and blow it completely out of proportion. Similarly, the little self has turned its forgetfulness into a powerful delusion. Because of its own irrational fear it has convinced itself the Divine is a volatile and vengeful force that can't be trusted. Although spiritual masters assure us that experiencing and reuniting with the Divine is life's most joyful and glorious purpose, the little self insists there is nothing more fearful than an encounter with Source. The sages testify that experiencing the Divine is the key to fearlessness; the little self insists Source must be feared. Ironically, we find ourselves drawn to something greater than ourselves, but remain terrified of finding it. As long as humans have searched for meaning and purpose, this paradox has played a significant role. Why? The influence of the little self cannot be avoided as long as we continue to desire separation and specialness. And that realization is not possible until we discern that separation is not working.

As we explained in Part Two, the little self (aka false mind or false self) and the ego are not the same. The ego is a necessary and natural part of the physical body and disappears as each body dies, but the little self is immortal. Although it remains oblivious of its indivisible connection with Source, it remains intact and aware as it projects one lifetime after another. The little self can hang onto the information it deems important from each lifetime, but it seldom bothers to remember the details of each lifetime it has projected. This is especially so in cultures that don't believe in reincarnation. Our current lifetime feels exceedingly important while we're projecting it, so why doesn't the little self bother to remember the details? Because the hundreds or thousands of lifetimes the little self projects hold little importance beyond the fact that they sustained the little self and its dualistic agenda. But at times, another lifetime is clearly remembered and memories or talents and preferences may carry from one lifetime to another. Although we might say that the little self has a 'warped' outlook, it still remains pure consciousness, and it isn't limited to projecting this material world. Between lifetimes, it can project whatever experience it can imagine. Many try to explain away the stories told by those who have had a 'near death experience,' but when we consider that consciousness never dies and is capable of projecting a completely convincing virtual reality of anything it imagines, the stories become feasible.

No matter how glorious they appear to be, near death experiences, experiences of life on heavenly planes, reuniting with dead loved ones or stories of astral travel in other realms all originate from the little self. How can we know that? These projections can vary wildly, but there is always one component that remains constant in all of them: separation and specialness. Because the little self was born out of a desire that requires a dualistic thought system to survive, everything it projects must be based on a dualistic foundation. Since that's the case, these illusions are not a return to the Reality of Oneness. As Plotinus explained, "True waking is not of the body, but from the body. Anything else is just a passage from sleep to sleep." Virtually all of these 'out of the world' projections still incorporate some form of separate body and the same old way of thinking Any paranormal experiences that are based on duality, separation and specialness are the product of the little self. As long as the little self remains in control, everything we experience remains part of our projected experiment.

When we decide we want to transcend separation, the barrier we erected between the little self and true Self begins to evaporate. The little self can and does struggle against Oneness. It often uses the brain and body as allies to keep us attached to duality. But remember, the little self is *not* a separate entity that can force you to do anything against your will because it is your will. *Since its part of your own consciousness, the little self has only the power that you choose to give it.* Rejecting separation and specialness and embracing Oneness starves the little self. Like a plant without the sun or rain that nourished it, it again becomes a part of the soil it came from.

The *Upanishads* are especially clear on the difference between the little self and the true

Self. We've looked at some of these verses before to illustrate the fact that we are not the body, but let's look at them again while keeping the difference between the self and Self in mind. We've included a few quotes from additional sages that illustrate this difference as well:

Renunciation is renunciation of the self—not of life. The end...is to know the Self...to realize your identity... The Self is one. Without the Self, there is no life.—*Isha Upanishad*

There are two selves, the apparent self and real Self. Of these it is the real Self...who must be felt as truly existing...There is no one but the Self. He who sees multiplicity but not the one indivisible Self must wander on from death to death...The all-knowing Self was never born, nor will it die. Beyond cause and effect, the Self is eternal and immutable...The truth of the Self cannot come through one who has not realized that he is the Self. The intellect cannot reveal the Self—*Katha Upanishad*

Self is everywhere, shining forth from all beings, vaster than the vast, subtler than the most subtle...He who finds the Self is free; he has found himself, he has solved the great riddle; his heart is forever at peace ...To become one with the Self is the only wisdom—*Mundaka Upanishad*

The Self is realized in a higher state of consciousness when you have broken through the wrong identification that you are the body, subject to birth and death. To be the Self is to go beyond death.—*Kena Upanishad*

To know the world you forget the Self. To know the Self, you forget the world—Nisargadatta

The real meaning of crucifixion is to crucify the false self, that the true Self may rise—Hazrat Inayat Khan (Brought Sufi beliefs, focused on Oneness, love and harmony, to the West 1882-1927)

Our true Self is the ultimate reality of the universe—Nishida Kitaro

You are Self—the Solitary Witness. You are perfect, all-pervading, One—*Ashtavakra Gita* (like the *Upanishads* writers, the author of this ancient manuscript is anonymous)

II. Soul or Spirit?

If you're used to the model of soul and spirit taught by most religions, our discussion of the self and Self may have sounded strange to you. Since the concept of Self (and quantum physics)

denies that you have ever been, or ever could be, a body, you may be wondering where the soul or spirit comes in. After all, you have probably run across both words in many spiritual writings, but we appear to be ignoring them. Language is fluid and constantly changing; so it is quite possible to read a text written by a spiritual sage who clearly understood the concept of self/Self, but used the words soul or spirit, to signify the Self. In that case, it's usually obvious from the remainder of the text what the sage meant. Take, for example, this quote from the *Isha Upanishad*, "To the illumined *soul*, the *Self* is all. For he who sees everywhere oneness, how can there be delusion or grief." [Italics ours] Here the writer used both the words soul and Self interchangeably, but as we know from reading the entire *Isha Upanishad*, the writer clearly understood and taught that the self/Self is pure consciousness. Also, please keep in mind when you're reading spiritual material that followers have often changed a master's words because they could not comprehend the meaning or had their own agenda. Or they may have mistranslated or put the words in common vernacular, which did not represent the original meaning.

The words soul and spirit can be extremely confusing to the spiritual seeker since most religions use the words, but have no understanding of the concept of self/Self. Since time, location and perception all have a powerful effect on language; let's take a closer look at how most religions use the words soul and spirit so we can get a clear understanding of how those views have shaped common misperceptions that cannot harmonize with our quantum Reality.

We must begin by acknowledging that few religions agree on the meanings of soul and spirit. However, most religions do base their understanding of soul and spirit on an outside god that gives and takes life. Some religions claim the soul is eternal and exists in heaven until it enters a body at birth and leaves it at death, although few explain what the soul is doing while it waits to come to earth. This definition is also similar to the New Age teaching, "you're not a body having a spiritual experience; you're a spirit having a human experience." Most religions who teach an eternal soul also believe that as a human, you must meet certain requirements before your soul is allowed to return to heaven. If you don't measure up, the soul will either be destroyed or go to a place of eternal punishment. In other words, the choices of the 'imperfect human' determine the soul's outcome, not the choices of the soul itself. Or, if the soul is supposedly making the decisions, it's hampered by the weak human form that imprisons it.

Other religions believe the soul comes into being along with the body and is given the opportunity to earn immortality. In all of these cases, the soul is believed to exist inside the body and animate it. In the first case the soul is considered immortal but may experience eternal bliss or eternal torment. In the second, immortality is conditional and by no means guaranteed. In some cases, religions use the words soul and spirit interchangeably, in others; the soul is a supernatural being while the spirit is the breath or spark of life God uses to animate the body. Some religions believe that the body and soul are one and the same, and claim the literal body/soul will be resurrected during a period of judgment. At that time, the restored body will either be given immortal fleshly life or be eternally annihilated. With all this confusion, how can we know what's true?

Most of us have seen the small pieces of litmus paper used to test whether a solution is base or acid, but the term 'litmus test' also means "any test that uses a single indicator to prompt a decision." Science has often been just that sort of litmus test when it comes to religious teachings. As we know, the Catholic Church had long claimed that God created the earth in seven literal days and placed it at the center of the universe. When science proved otherwise, the church was forced to 'reinterpret' its supposedly inviolable teachings. Since quantum physics has exposed us to a more accurate picture of the workings of the universe than we've ever had before, the 'quantum litmus test' is a decisive way to decide whether religious teachings correlate with universal truth or not. If a teaching fits into the quantum paradigm it's worth our consideration, if it can't, we would be justified in dropping it even though it's part of a 'sacred text.' This is certainly true when it comes to teachings that define who and what we actually are.

Even though it appears that the body dies, it's a scientific fact that energy can appear as matter and matter can return to energy, but the whole remains, and nothing 'dies.' This truth speaks to the eternal nature of the Oneness of All That Is. The quantum model supports the sage's realization that everything exists within Oneness and shares immortal life with Source. We are the Self of pure consciousness, the very 'stuff' of Ultimate Reality, infinite and eternal. As the ancient *Bhagavad Gita* testifies, "There has never been a time when you and I...have not existed, nor will there be a time when we will cease to exist." We don't have to worry about our life ending any more than we would be concerned about Ultimate Reality disappearing. Even if our dualistic system destroys life on earth or the material universe implodes sometime in the next several billion years, it will not affect the One Mind we all share. Some scientists have likened the expansion and possible contraction of the material portion of the universe to a 'divine breath.' If that's true, we have no idea how many times a material universe may have already expanded and contracted, or how many more times that may happen. Regardless, we're assured that we will all play a part in the adventure.

Since our true Self *is* the non-local consciousness that projects the material universe, it would be as impossible for the Self to be trapped within a body as it would be for you to hold the ocean in a paper cup. The *Ashtavakra Gita* explained it this way, "I am wonderful indeed...The universe appears within me but I do not touch it." Nisargadatta agreed, observing "When I say 'I am'...I mean the totality of being, the ocean of consciousness, the entire universe of all that is and knows." Rumi pointed out, "Everything in the universe is within you...Why do you weep? The Source is within you and this whole world is springing up from It." And renowned physicist Albert Einstein recognized that we are far more than the body when he said, "A human being is part of the whole, called by us universe...he experiences himself, his thoughts and feelings, as something separated from the rest—a kind of optical delusion of consciousness."

Since the self/Self exists outside of time and space, the little self can experiment with separation and specialness over and over again, projecting as many lifetimes as it wishes in its attempt to prove its dualistic system can surpass Oneness. Since time/space is a construct of the material world that we project, there are no 'time limits' on our experiment. When you think about it from that perspective, it would be unrealistic to think that we could fully carry out our experiment during one limited human lifetime. The many lifetimes we've projected are similar to buying a lottery ticket every week. If you don't win this week you may win next week. If the little self didn't get the results it wanted in this lifetime, it can just keep playing and betting that the next lifetime will be the winner. By now, we've probably each experienced virtually every time period, gender, race, ethnicity, nationality, sexual preference and station in life. We've been rich and poor, healthy and sick, cruel and kind, powerful and weak, saint or monster. When Buddha 'woke up' from the dream of this world, he said that thousands of his previous lifetimes flashed before him. If we instantly remembered every lifetime we've experienced, it would be blatantly obvious that duality can never successfully rival Oneness. But as long as the little self projects each new lifetime with the unshakable belief that it will finally win the specialness lottery or that duality can successfully rival oneness, the experiment will continue.

Although most of us don't remember past lives, researchers have amassed and authenticated thousands of verified testimonies from people who recall past lifetimes. This phenomenon is usually associated with reincarnation, the belief that a soul enters and inhabits one body after another as it tries to work its way off of the wheel of karma. As we've learned, consciousness *projects bodies* instead of inhabiting them, and there is no supernatural scorekeeper dealing out appropriate rewards and punishments required by the karmic system. Because each body is projected from consciousness, it's not the body remembering a past life but a memory retained by little self. As we've said before, the personality is the result of collected information and conditioning. We cling to the personality and believe it's who we are, but it actually is a product of the brain and has nothing to do with our Reality. As Shankara realized, "The mind of the experiencer creates all the objects which he experiences while in the waking or the dreaming state," which includes the body and personality The personality that's held to be so precious during each human life is not a part of the consciousness of the little self that projects it. In fact, the little self cares little about the details of each personality. But as we've discussed earlier, it can preserve information from each lifetime that it considers valuable and make use of it during the successive lifetimes it projects. During its entire existence, the little self has had the opportunity to either realize that our experiment isn't working, or pursue new tricks to try the next time around. *If it weren't for free will and choice, the little self would be stuck forever in the cycle of projecting an endless stream of bodies; creating personalities for them and watching them die.*

III. Experiencing All That Is

As we've said several times, and will continue to say, *personally experiencing the Divine is the real and only teacher.* The ancient Buddhist *Lankavatara Sutra* testifies, "These teachings...are not the Truth itself, which can only be self-realized within one's own deepest consciousness" Only though this direct experience can you *know* who and what you truly are. The words of sages who have returned to Self can encourage you and point you in the right direction, but they are no substitute. The 16th century German mystic Jacob Bohme explained, "Spiritual knowledge cannot be communicated from one intellect to another, but must be sought for in the spirit of God." In other words, it's the Self that knows the Divine, not the little self, brain or body. The door must be opened between self and Self before experience takes place, and the door is opened by our willingness to let go of our own attachments and aversions, by opening our heart to whatever Ultimate Reality wants to show us. Our personal experience demonstrates that it's possible to go from the consciousness of the little self to the Self in an instant, but sages tell us that the shift occurs more commonly when we begin by letting go of the misperceptions that fuel the little self one by one. (The experience of an instant shift from self to Self does not make anyone special, they still have a lot of work to do to 'catch up' with their instantaneous knowing.)

In his book *The Perennial Philosophy*, Aldous Huxley outlined several fundamental qualities that have been common to successful spiritual seekers. These qualities have nothing to do with secrets, formulas, methods or spiritual practices, but can best be understood as a particular mindset. A mindset is a set of choices we make, not something we can mimic, learn or do. Choices are free and available to everyone no matter what our personal circumstances might be. Let us repeat this important component to spiritual understanding: *A mindset is a set of choices we make, not something we can mimic, learn or do. Choices are free and available to everyone no matter what our personal circumstances might be.*

One of the factors present in the mindset of successful seekers is known by a variety of terms such as 'empty hands,' 'poor in spirit' or 'beginner's mind.' These terms picture our willingness to let go of our misperceptions, which includes our conditioning, attachments and aversions. When we're trying to prove our preexisting beliefs, our hands are symbolically full and we feel certain that our spirit is already rich. Either way, there is no room for anything new. In Zen, the expert's mind is certain, while the beginner's mind is open to new possibilities. Huxley also mentions being 'pure in heart.' This doesn't mean that we need to 'clean up our act,' purify or perfect ourselves before approaching the Divine. On the contrary, it symbolizes a heart that is not looking for things, but desires a deep, personal relationship with Ultimate Reality for the sheer joy and freedom that's inherent in the relationship. Ansari of Herat, an early Sufi master demonstrated a pure heart when he prayed, "Each one wants something which he asks of You, I come to ask You to give me Yourself." Jesus' pure heart was also captured in the words, "Man cannot live on bread alone, but by every word that proceeds from the mouth of God." Here are a few additional quotes that reinforce the importance of examining our motives and any thoughts we may be clinging to:

In its true state, mind is naked, immaculate; not made of anything...clear, without duality...timeless ...not realizable as a separate thing, but as the unity of all things. The One Mind is Total Reality—Padmasambhava, aka Guru Rinpoche, Lopon Rinpoche or Padum, 8th century Buddhist, considered by some to be a second Buddha

The truly great man dwells on what is real and not what is on the surface, on the fruit, not on the flower...When I let go of what I am, I become what I might be...Can you step back from your own mind and thus understand all things? Can you coax your mind from its wandering and keep to the original oneness? Can you cleanse your inner vision until you see nothing but the light?—Lao Tzu

If the doors of perception were cleansed, everything would appear as it is—William Blake

Through an uncompromising, absolute and pure detachment from yourself...you will be led...to that radiance which is beyond all being—Pseudo-Dionysius the Areopagite, 5th or 6th century mystic Christian

As long as you have certain desires about how it ought to be, you can't see how it is—Ram Dass,(born Richard Alpert, 1931) contemporary American spiritual teacher

In the unitive experience, every trace of separateness disappears; life is a seamless whole...The experience of unity must be repeated over and over until the seeds [of separation] are burned out and can't sprout again—Eknath Easwaran

Mastery lies not merely in stilling the mind, but in directing it towards whatever point you desire... As soon as...clouds of illusion are scattered, that which man now begins to see is nothing but the truth which has been there all the time—Hazrat Inayat Khan

After life, free will and choice are the most loving and generous gifts Source has given you. In this world, acquisition is our primary choice. To return to Oneness, we choose to let things go. As the 9th century Zen master, Huang Po advised, "Just discard all you have acquired...rid yourself of the whole gamut of dualistic concepts." Meister Eckhart understood that, "God is not found in the soul by adding anything but by process of subtraction." This doesn't mean that we must renounce our material possessions, just let go of our attachment to it. When Sosan said, "Don't keep searching for the truth; just let go of your opinions," he realized that everything false is 'out there' but truth hides within us under layers of mental debris. Many sages refer to this debris as 'veils' because they obscure our spiritual vision and leave us with only the limited sense perception of the body. As we've learned from quantum physics, the senses connect us to virtual reality rather than Reality.

The words of the sages that we've quoted so far were correct in telling us to let go of our misperceptions, but the brain is not a vacuum. Unless we offer it something more valuable to replace what we let go of, it will just keep going back to its old way of thinking. Although our goal is eliminating the wall between self and Self, we need some cooperation from the brain. For this reason, the brain is an extremely effective 'gatekeeper' for the little self. Let's confront some of the issues that keep the brain frozen in fear. If we understand what's at the core, the brain will see the wisdom in allowing us to swing the gate between self and Self a little wider. One of the world's greatest purveyors of misperception and fear is religion, so we'll start there.

IV. Religion, a Tale of Fear

Religion usually keeps us from spiritual discovery for one of two very powerful reasons: we're either afraid to let go of it, or we're so disgusted by it, we throw the spiritual baby out with the religious bathwater. In the first case, we've usually been taught that we are either unable or unworthy of finding God on our own, or that something terrible will happen to us if we leave the church to pursue our own spiritual path. In the second case, we mistakenly believe God and religion are synonymous and condemn them both in one breath. But in both cases, it's actually our misperceptions about the Divine that bring us to those conclusions. We've already addressed some of the issues involved in each of these roadblocks, but there are several more worthy of our attention. We'll begin with some of the fears that can keep us tied to religion.

Fear of Authority

From ancient times, religion and politics have been locked in a symbiotic relationship that has strengthened both groups. In the past, it was common for priests and kings to hold the same degree of power, and even now, religion remains intimate with the state and holds a great deal of the world's wealth. Even in countries where church and state are supposed to be separate, it's common to see the clergy called in to confer on important issues or 'bless' official government functions and encourage their followers to fight the state's wars. The Emperor Napoleon observed, "I'm surrounded by priests who repeat their kingdom is not of this world and yet they lay their hands on everything they can get." However, he also believed, "Religion has always had one very useful tool—it keeps the poor from killing the rich." The 1st century Roman philosopher Seneca concurred by saying, "Religion is regarded by the common people as true, by the wise as false, and by the rulers as useful."

Considering religion's usefulness to the state, it should come as no surprise that society conditions us to believe that religion is the only route to God. The fact that many religions have a long history, sacred books, great wealth, impressive buildings, priceless art, beautiful music, and the pageantry of 'holy days' adds to the conviction that "might makes right." This is an extremely powerful message, both overt and covert, that begins to infiltrate our brain very early in life. Most religions are authoritarian by nature, and are often associated with other important authority figures in our lives such as parents, relatives or teachers. As infants, we quickly learn to recognize that our life and well-being depends on people who have power over us. As we grow up, this concept is regularly reinforced. We're frequently reminded in hundreds of little ways that if we want to get along in life, it's best to cooperate with the 'powers that be.' While that may be the case in childhood, such thinking stifles us as adults. Nonetheless, those in power rely on our buying into this view and cooperating with the program they've outlined for us. But as a scientist discovering radical new concepts Albert Einstein remarked, "Unthinking respect for authority is the greatest enemy of truth." Regardless of the clout wielded by the powerful combination of church and state, each and every seeker who has experienced the Divine has recognized the need to see beyond the standard message they promote. As Joseph Chilton Pearce pointed out, "The good news of Jesus was at a radical discontinuity with the...culture of that time." Jesus was a rebel who spoke openly against the religious institution that ruled his society, and he would do the same today.

There is no doubt that it takes courage to step away from the accepted ideals of society, but as Hafiz so clearly put it, "No one but a rebel can get their mitts on God. At some point you will have to wean yourself from the pack...We have not come here to...confine our wondrous spirits but to experience ever more deeply our divine courage, freedom and light...Come, join the courageous who have no choice but to bet their entire world that indeed, indeed, God is Real." Nisargadatta explained why spiritual courage can seem so challenging, "The world in which you live has not been projected on you, but by you...To know the world you forget the Self. To know the Self, you forget the world...The search for Reality is the most dangerous of all undertakings, for it destroys the world in which you live." One of the most important factors in understanding Reality is realizing, and accepting, that it can't be found in this world.

Whether we're aware of it or not, countless times each day we decide whether to continue as part of the pack that's pursuing separation and specialness, or to make a choice for Oneness. But as 20th century spiritual philosopher Jiddu Krishnamurti pointed out, "We have created the religions, the beliefs, the dogmas, out of fear. It is in that society that you live...Society merely cultivates the known...To find the unknown, it is essential to be free of society." Rumi explained that the authority figures of this world are certain of their own importance and the rightness of their beliefs but, "They believe they rule the world, unaware that what they rule is a heap of ash that one breath of the Beloved could disperse forever...I know this world; it never fulfills its promises. What an absurd fate to give your life to this world...Your task? Escape...the fires of madness, illusion, and confusion that are, and always will be, the world...I'm tired of cowards, I want to live with lions." *If we want to know the Divine, it takes a lion's courage, but as we discover the Self, we also realize we already have all the courage needed for the task.*

Fear of Family and Friends

Deciding we no longer want to buy what society is selling can be done quietly. We have no obligation to advertise what we're doing or explain ourselves to anyone we know. But many of us come from families that have been associated with a particular religion for generations. Family pressure and expectations keep many people locked in religion even though they don't find it satisfying. We've often heard people say, "If it was good enough for my grandfather, it's good enough for me," or "I was born a _____, I'll die a _____." (Fill in the blanks with the religion of your choice.) Both these statements boil down to loyalty to family as much, or perhaps more, than to the religion itself. Some feel that leaving a religion is tantamount to accusing their family of being wrong or stupid. While they may not care for religion, they care a great deal about their family. And for many, the church doubles as a social center and support system where they feel they can safely associate with others who have similar values.

We understand these fears particularly well since we realized ahead of time that leaving the church would also mean losing family, lifelong friends and the support system of the church we had been involved with for so long. But as painful as such a change can be we found that staying where we were and living a life of pretense was even more destructive to ourselves and the ones we love. Feeling one thing and living another sets up an internal conflict that inevitably harms both the body and the mind. Siddhartha Gautama, a prince later known as the Buddha, left his wealth, his position as prince and his family to pursue truth. From the world's perspective, he gave up a great deal, but he observed. "The wise, knowing what is trivial and what is vital, set their thoughts on the supreme goal." In the *Dhammapada*, a text attributed to Buddha, he advised, "If you find no one to support you on the spiritual path, walk alone. If you cannot find a friend who is good, wise and loving, walk alone. It is better to be alone than live with the immature." *The only way we can influence anyone else is through our own example.* Like Siddhartha Gautama, the greatest gift we can give to the world is our return to Self.

Fear of Unworthiness

Although we're taught to see the clergy as 'spiritual helpers,' the division between clergy and laity also creates a rift. A hierarchy of trained clergy emphasizes the notion that only a very few are specially called to have a relationship with God. The clergy emphasize the fact that they are shepherds who will protect and lead, while the duty of the 'flock' is to remain as docile and submissive as sheep. The word 'clergy' originally meant 'learned professionals' and 'laity' meant 'unlearned amateurs.' In the early days of Christianity, these labels were an accurate description. Only a small percentage of the population could read since access to education or written materials was extremely limited. Education, religious training and access 'holy books' created and maintained a wide gap between clergy and laity. Centuries later, even though education and written materials are within the reach of nearly everyone, church goers are still conditioned to believe they don't have adequate knowledge or the understanding necessary to question concepts that may make little or no sense to them. When they do have the courage to ask questions, they're often told that God is a mystery only the specially chosen can comprehend. Sadly, most would rather swallow this lie than make the effort needed to discover the truth for themselves.

Instead of encouraging the laity to recognize their unbreakable connection with Source, parishioners are regularly reminded of teachings that emphasize their sinfulness and humanity's inherent unworthiness to approach Source directly. But as author Tom Robbins pointed out, "It is not only that religion is divisive and oppressive, it is also a denial of all that is divine in people." And as Catherine of Siena, a Catholic mystic observed, "Strange that a priest would rob us of knowledge and then empower himself with the ability to make holy what already was... Lose yourself wholly; and the more you lose, the more you will find." As Catherine of Siena implied, the entire concept of sin, and in turn unworthiness, was manufactured by the little self.

What do you think of when you hear the word sin? Most religions define sin as 'falling short' or 'missing the mark,' but that's rather vague. One dictionary defines sin as "an offence against religious or moral laws," and another says sin is "a state in which the self is estranged from God." As we've learned, free will would be destroyed if the Divine interfered in our experiment and began issuing rules and regulations for us to follow. The 17th century philosopher Baruch Spinoza pointed out that the concept of sin did not originate with Source when he said, "Sin cannot be conceived in a natural state, but only in a civil state, where it is decreed by common consent what is good or bad." The gnostic *Gospel of Philip* joyfully explains, "He who has knowledge of the truth is a free man...Those who think that sinning does not apply to them are called free." And Hafiz adds, "Forget every idea of right and wrong any classroom ever taught you...God...said: 'There is no man on this earth who needs a pardon from Me for there is really no such thing as sin.'" Since we're the ones who have invented all religious and/or moral laws, we are also justified in ignoring or rescinding them. Convincing ourselves that rules or moral standards were dictated by God cannot make it so. Source has no interest in whether we live by rules and standards or not, the only thing that interests Source is whether we are ready to return to oneness.

Sadly, as pointed out by Joseph Chilton Pearce, "The...accusation of sin is part of the very fabric of our culture...the more subtle its presence, the more powerful its effect." Many religions cling to the assumption that all humans are estranged from God because we're all inherently corrupt and in need of redemption regardless of our behavior. If this were true, it would certainly be a sad commentary on God's creative ability! The concept of 'original sin' (sin inherited from Adam and Eve) came about because we forgot that Divine Presence puts no limits on our projections. Because we think in dualistic terms, we have labeled some things as good and others as bad. We're not advocating a life of reckless abandon simply because every unloving thing that we think or do can take us deeper into our dream. Still, there's nothing we could project that would cause Source to label us 'sinners,' condemn us or sever our connection. And at times, a horrendous act has served as a wake-up call. We don't sin, we misperceive. It's our misperception of who and what we truly are that's at the root of our misery. Misperception can be corrected, and when it is, the pain it caused disappears with the thoughts that caused it. Since we always have been, and always will be, one with the Divine, our worthiness is a given.

Throughout human history fear has been a crucial tool used to enslave minds. When we're threatened with eternal damnation for questioning dogma or taking responsibility for our own spirituality, it takes a great deal of courage to put the threat to the test, especially when we feel unworthy. As we've repeated many times, the brain prefers the status quo. The brain feels far safer as part of a group, especially one that has the favor of authority figures. It is undoubtedly far easier to hand over responsibility for your spiritual well-being to someone else, but religious leaders are playing a fantasy role that the Divine does not recognize. As Rumi explained, "There is nothing outside yourself, look within. Everything you want is there. Remember, the entrance door to the sanctuary is within you... Knock on the inner door. No other will take you to joy." Sages who have mastered the false self all testify that they found the Divine within. Here are a few more of their words:

If you look for the truth outside yourself, it gets farther and farther away—Tung-Shan

Open the door for yourself that you may know what is...Whatever you will open for yourself, you will open—*Dialogue of the Savior*

Stop searching for God outside yourself. Look for him within. Thus you will find in yourself a way out of yourself— Monoimus

It's true that the pomp, ceremony, rituals and history of religion can engender a sense of awe and the comforting feeling that we're associated with something far bigger than ourselves. Many believe that a religion's long history, wealth and power are enough to prove it has God's approval, but many religions that utterly disagree with each other share those qualifications. As author Tom Robbins put it, "Religion is nothing but institutionalized mysticism. The catch is, mysticism does not lend itself to institutionalization. The minute we try to organize mysticism, we destroy its essence." It is, after all, impossible to organize personal experience; nonetheless, even some who claim to understand gnosis have still tried to create a religion around it, complete with rules and hierarchy. At that point, it becomes an 'ism' and can no longer fit the definition of gnosis. Although the word Gnosticism is commonly used, it is completely out of sync with the definition of gnosis. When the suffix 'ism' is added to a word, that word then describes a distinctive doctrine, theory, system, or practice such as Catholicism, Judaism or Buddhism. But the individual experience central to gnosis makes gnosis impossible to organize, let alone institutionalize. As we said earlier, willingness to let go of the brain's preconceived notions and accept the experience without reservation makes gnosis possible, but there are no practices, systems, rules or doctrines involved. When can find no way around using the word gnosticism, we're referring to a personal, organic spiritual approach that has been used by seekers in all cultures, eras and areas of the globe.

Religion reduces the Divine power and glory we want to experience to the pathetic shadow humans can produce. Lahiri Mahayana a 19th century Indian sage advised, "Clear your mind of dogmatic theological debris; let in the fresh healing waters of direct perception." Yes, it can feel frightening to walk away from the pseudo-security offered by religion or stand up to its seemingly ubiquitous power, but nothing man projects has the ability to overpower the Reality and innate worthiness of Self.

Doubt

Sages continually remind us that Ultimate Reality not only can be known, but wants to be known. If we doubt and question during the process of knowing, that is perfectly acceptable, in fact it's absolutely necessary. As the Buddha so wisely taught, "Do not believe in anything simply because you have heard it. Do not believe in traditions because they have been handed down for many generations. Do not believe in anything because it is spoken and rumored by many. Do not believe in anything simply because it is found written in your religious books. Do not believe anything merely on the authority of your teachers and elders. But after observation and analysis, when you find that anything agrees with reason, and is conducive to the good and benefit of one and all, then accept it and live up to it."

The last sentence of Buddha's advice is especially important. Truth is, without exception, beneficial for all! That is because truth harmonizes with universal Oneness. This is another 'litmus test' we can use when we question something we've heard or experienced: is it conducive to the good and benefit of one and all? As our quantum model demonstrates, Ultimate Reality does not consist of 'separate parts.' The road back to Oneness is not found in anything that promotes divisiveness and separation, but in everything that moves us closer to love, peace and harmony with All That Is. Doubt is your enemy if it keeps you locked in fear, your friend if it builds your courage to question and change.

Disgust

For many, it's not their religious affiliations that hold them back from spiritual discovery, but their disgust with the track record of religion. In Matthew 7:16, Jesus gave his followers another litmus test when he said, "By their fruits you will know them." When we look at the 'fruit' that religion has produced, much of it is decidedly rotten. Since most religions preach love and peace, it's reasonable to expect that they should be the outstanding force for good in the world, not an instigator of intolerance, hatred and violence. But rather than take the lead as peacemakers, religion has been, and continues to be, the cause of innumerable wars as well as encouraging, supporting and blessing countless others. Jesus is labeled "the prince of peace" but many of Christianity's staunchest supporters give their allegiance to the gods of war. It's also impossible to know how many have been tortured or died as a result of forced conversions or religious persecution (both inside and outside churches) or to measure the amount of emotional suffering religion has caused.

Author Joseph Chilton Pearce properly asks these pointed questions, "What are the actual, tangible results of the lofty religious institutions that we have known throughout history? Why, after two thousand years of Bible quoting...has civilization grown more violent and efficient in mass murder?" Pearce's questions could just as easily be asked of most non-Christian religions as well. If we believe that God is the originator and supporter of religion, it would be logical to blame God for everything done in the name of religion. If we're disgusted by religion, we could not help but find God disgusting. It's true that many religious organizations do good works, but often recipients are expected to 'pay' for the necessities they receive with conversion to the faith. Regardless, our quantum model of the universe does not support a link between Ultimate Reality and religion. Disgust with religion is a valid reason for dropping religion, but instead of dropping Source, we can look elsewhere. After asking his scathing questions about religion, J. C. Pearce also observed that

"Spiritual transcendence and religion have little in common…these two have been fundamental antagonists."

Even when we can separate God from religion, many remain disgusted by the mere thought of God. They see the evil in the world and conclude that God could not be good. They look at the misery and suffering that has plagued human history and feel certain that a God who has the power to stop it and doesn't, deserves their contempt. *If this world was real,* those views would certainly be valid and justified. But as quantum physics demonstrates, this world is a perfectly safe virtual reality, a projection from consciousness at the quantum level. And as the sages tell us; these projections come from the conflicted mind of the little self. The world we see is a result of a collective projection; still, we're each responsible. The problem lies in the fact that we've become so deeply involved in our dream, we've forgotten this truth. Because the pain feels very real to us, we might still feel that Source should step in and stop our experiment. Yes, Ultimate Reality gave us a safe venue to play out our desire for separation and specialness. And yes, when we feel that we've had enough, Source will aid in our escape. But it would violate free will, render our experiment invalid and keep our challenge forever in question if Ultimate Reality interfered in our world. It's rather like a basketball game where a coach can cheer on the team and give advice to players when they're on the sidelines, but is not allowed to come onto the floor to interfere with the game.

Holding a grudge against the Divine based on what we think Source should be doing won't teach us anything about the true nature of the Divine. For most of us, what we've learned about God came to us from other people who didn't understand Source any better than we do. As we said earlier, when a spiritual seeker experiences the Divine directly, others often make a religion around that person rather than seeing the need to have the experience themselves. A spiritual master might write down what they've learned, but it can still be misinterpreted. At times someone records the words of the sage exactly as they were said and meant, but far more often a writer, blinded by their own conditioning, misunderstands and modifies what they heard. Many 'sacred texts' have been worked over by those with an agenda to the point that the original message has been lost. On the other hand, some texts still contain nuggets of truth. How can we tell the difference?

Once we've experienced the Divine the difference will quickly become evident. Until then, we can begin by doing as Buddha suggested and let go of any teaching that would not benefit everything in existence.

The bottom line? The only way to know Source is to experience Source. When you want to understand someone, you don't rely on what others tell you about them. And you wouldn't want anyone to think they knew you based strictly on the opinions of others, especially if those talking about you hadn't met you themselves or disliked you. Still, we seem to think that we can understand the Divine from reading words that have passed through hundreds of hands before we've seen them, or by listening to someone else who has never experienced the Divine. It's as if Source is being condemned on the basis of hearsay, something that's inadmissible in a court of law. We can't claim that the Divine is inaccessible, because quantum physics demonstrates that we are one with Source. When we experience that oneness for ourselves, the opinions of others become meaningless. As the 8th century Sufi mystic Rabia explained, "When I entered God, my vision became like His, it flooded out over existence. I knew no limits." And Rumi joyfully tells us, "When you eventually see through the veils to how things really are, you will keep saying...this is certainly not like we thought it was."

It's important to remember that the religions we create, the churches we build, the books we write and call sacred, the doctrines, rules and 'moral standards' we manufacture never have been, and never will be, anything other than our own inventions. The great good we do as well as the terrible atrocities we commit are *all* the result of our own drive for separation and specialness; the Divine is not responsible. While religions *tell* us who they think God is; on the spiritual path we empty ourselves of preconceived notions so the Divine can *show* us what actually is.

We do see evil in this world, but again, it is not something Source set in motion. Both Jesus and Buddha were said to have been tempted by evil entities who offered them great power before they began their public teaching, but it's important to remember that symbolism is an integral feature of ancient texts. When we take these symbols literally, we can quickly get confused. Jesus clarified what was meant in Mark 7: 20-23 when he explained that evil is not something that originates outside of us but "issues from our own mind." Buddha also made this clear when he said, "All that we are is the result of our thoughts. We are made of our thoughts; we are molded by our thoughts... Your worst enemy cannot harm you as much as your own unguarded thoughts... It is a man's own mind, not his enemy or foe that lures him to evil ways." Instead of being tempted by something outside them, Jesus and Buddha both realized that temptation came from within. Instead of being someone or something that lures us or tricks us into wrongdoing, *temptation is best defined as anything we allow to entice us into putting our illusionary projections before our true Self, separation before Oneness.*

As Jesus explained, 'evil' is one of the inevitable outcomes of our desire to experiment with separation and specialness. As we've said before, duality was a necessary foundation for our experiment. If everyone is special, special no longer exists, so 'not special' was necessary. In duality, what we label as 'good' will always be accompanied by what we choose to call 'evil.' The pursuit of specialness throws the harmony of Oneness out of balance and actually creates more of what we don't want rather than what we do want. You can understand this imbalance if you visualize an artist's gray scale. Imagine a dab of white paint at one end of a piece of paper and a dab of black paint at the other. Now visualize the space in between the black and white paint as varying shades of gray that get progressively darker toward the black paint, and progressively lighter toward the white. If you labeled the white dab of paint 'special,' then the far greater part of the scale (all the gray and the black) must be 'not special.' This makes specialness very scarce and turns life into a competition. This desperate competition drives many to feel sure the only way they can experience specialness is through the use of means that are labeled 'evil.' On the other hand, in Oneness each of us is unique. Since everyone can be unique, there is no opposite or duality involved.

Free will allows us to play out even the most destructive desires; if it didn't, it wouldn't be free will. During the hundreds, or even thousands, of lifetimes we've each experienced, it's highly probable that we've done things that we would currently consider unconscionable. Knowing this, and that evil can't exist in Reality, helps us to realize there's no point in judging what others are projecting. No matter how horrific, evil is nothing more than thought, and there are no evil entities that can test, tempt, control or harm us. There is no 'cosmic' war between good and evil, and no judgment against us for what we project. The question was never whether Jesus or Buddha would remain faithful to God or choose to serve an evil entity. The paramount issue for them was whether or not they would continue to project virtual reality or turn their backs on it and return to Oneness. Whether we're aware of it or not, that is the question that's paramount in our lives as well.

We truly do live within our own image of this world. Since it's our collective thoughts alone that manifest virtual reality, there's no escaping the fact that we're all responsible for whatever is experienced in this world. Many feel offended by that assertion, arguing that they have only positive thoughts toward others. True as that may feel to us, we wouldn't be projecting this lifetime if we hadn't wanted to continue pursuing separation and specialness. The controversial 18th century theologian William Law was correct when he said, "Men are not in hell because God is angry with them; they are in darkness because of what they have done to the light." That concept can feel very uncomfortable at first, but the fact that we're responsible is also the best possible news we could receive. Why? *Our thoughts are the one thing that we can control.* When we accept personal responsibility we're free to move past the erroneous belief that we are mere victims of an unfair and unjust world. Having choice and using it restores our personal power and gives us the freedom to make more positive choices. During each lifetime we have the opportunity to 'wake up,' see this world for what it is, and decide we've had enough. That has been true for every spiritual master and it's true for you!

V. Misperceiving Source

The 'Male' God

Even when we realize that religion is our own projection, some of the erroneous thoughts we've constructed about God are difficult to shake. Before going on to Part Four, we'll look at several of these concepts from the quantum perspective. One issue that's frustrated many women is the assumption that God is male. Some religions have included Goddesses, but they were often under the headship of, or the consort of, a more powerful male God. As a result, men have felt justified in wielding authority in God's stead and have that authority to rationalize demeaning and mistreating women. Ugly as their behavior has been, it's not surprising that in the race for specialness, discrimination and force are often employed. The Sufi master Hafiz recognized the problem when he said, "Let's call God a Her sometimes to balance out all that male testosterone that took over the world's holy books." But calling the Divine 'Her' or switching to goddess worship does little more than swing the pendulum in the opposite, yet equally unbalanced, direction.

The decision to label some qualities or characteristics as male and others as female is strictly a dualistic endeavor, as is the choice to imagine that Ultimate Reality exists in a male or female form. Divine Oneness transcends gender duality by containing all possible qualities as potentials. This means that Ultimate Reality, like the yin-yang symbol, contains ALL the qualities that we think of as male and female in perfect balance, but cannot be described as either one. The Yin and Yang are not the symbol of the necessity of opposites but rather the celebration of their interconnection and indivisibility in Oneness. As we learned earlier, spiritual sages identify Ultimate Reality as pure love, a wholly gender neutral quality. Because it's inaccurate to infer that the Divine is anything like a human or could possibly be one sex or the other, we make the effort to stick with gender neutral terms such as Source, the Divine, Ultimate Reality, All That Is, the One, etc.

Legalism

Authoritarianism has regularly been a hallmark of male dominated religions, and is virtually always accompanied by a plethora of laws and rules. In the case of religious authority, the focus is generally on 'moral values' that attempt to control the body and brain. Even though some of these values seem to make sense for society in general, such as edicts against violence, there are always those ready to break them. Thousands of years of religious history have made it clear that it's impossible to legislate morality or force anyone to love. If religious rules had any value outside of controlling us through fear, they would have made a positive difference by now. At best, rules keep us in line; but they can't 'improve' us because they don't reach the heart. Ironically, those who claim faultlessly 'moral' conduct regularly become profoundly immoral. Focus on being love and 'morality' takes care of itself. Become love and it will guide you to a far higher plane, one where rules and moral values are unnecessary. Unlike rules and laws that constantly need to be rewritten to cover unforeseen occurrences, love works in every circumstance. As William Law put it so well, "Love is infallible; it has no errors, for all errors are the want of love."

Even if we've never been affiliated with a religion, we may still think of God as an authoritarian rule maker. But as we've learned, Source is not involved in our free will experiment, and is certainly not sitting on a cloud using a rule book to keep track of what we do with the bodies we project. *All laws and rules serve the purpose of protecting and enforcing the thought system of the ones who made them.* In our dream, it is the little self, engrossed in the desire for specialness that creates rules as a means of controlling others. Many believe that God abhors homosexuality; but as we've learned, the Divine does not judge what bodies do. As far as any statements against homosexuality in the Old Testament of the *Bible* are concerned, there is an explanation that has nothing to do with sexual preference being right or wrong. The Jews were a tiny nation that desperately needed to grow its population as quickly as possible if they were to survive, so it's no surprise that they instituted laws to discourage homosexuality. They also supported polygamy and concubines, and when they conquered another nation, they killed all but the virgins who would be used to help them in their drive to reproduce. Frankly, current prejudice against another person's sexual preference or gender identification is no doubt based on groundless fear.

Religion has proved over and over again that it is extremely easy to convince people of their 'sinfulness' by handing out rules that are impossible to keep. Instead of helping the parishioners to freedom, the rules constantly condemn them. In the gnostic *Gospel of Mary*, Peter asked Jesus, "What is the sin of the world?" Jesus replied, "There is no such thing as a sin" and explained, "This is why you get sick and die; because you love what deceives you." How would you feel if the courts started prosecuting people for the 'crimes' they committed while playing a video game? Or what if actors were suddenly arrested for the crimes they acted out in movies, plays or during TV shows? No matter how heinous something *appears* to be in a video game, movie, play or TV show, we know that it's not real. Similarly, the Divine sees no reason to judge or punish what illusionary bodies do or bless them for how well they follow arbitrary man-made rules. Nor is Source concerned about judging or punishing the little self that projects our virtual reality. However, we would be wrong if we thought the Divine was completely uninterested in what the little self projects. What captures Source's attention is our waking up. As we let go of anything in us that is not love, the Divine pays attention. Jesus lived in a male dominated authoritarian society controlled by a seemingly endless array of religious laws that covered virtually every area of human life. Imagine the shock Jesus' follower must have felt when he told them that all these laws would be fulfilled if they loved God with their whole heart and loved their neighbor as they loved themselves. (Matthew 22:37-39) The 'law of love' Jesus recommended is not a rule you must follow, but a truth that describes how the universe actually operates. When you become love, you are in harmony with All That Is. Rumi explained, "There is a rope of light between your heart and [Source] that nothing can weaken or break...The essence of the whole matter is Love...whenever you detect love growing awake in you, feed it so it may open its eyes further...Love is infinite and so is the transforming power that streams from it...Open your heart and know: the expansion of the heart is infinite."

The Path of Suffering

By the time we're old enough to think seriously about God, we've already been conditioned by images and stories that convince us that hardship, suffering and sacrifice are an unavoidable part of spirituality. Judaism and Christianity are founded on the cornerstone of suffering, sacrifice and martyrdom. The *Bible's* Old Testament is filled with the suffering of God's chosen nation, which took place regardless of whether or not they were obedient to religious laws. If that were not enough, it is those who God supposedly held most dear, like Job, who were purposely tortured to prove their faith. In fact, the *Bible* book of Job tells us that God and Satan made a bet about Job's loyalty that resulted in the deaths of Job's entire family, the destruction of everything he owned, and the loss of his health and his friends. To top it off, God then allowed Satan free reign to test the rest of humanity using whatever evil means necessary. No wonder so many are willing to have a priest or minister act as a mediator between them and such a hateful God!

Orthodox Christianity is based on the belief that a 'loving' God offered his only son as a bloody sacrifice to pay for the sin of Adam and Eve. For every image of a smiling Jesus, there must be ten thousand that focus on his suffering. When the persecution of early Christians broke out in the first century, many felt it was their duty to follow in Jesus' footsteps and offer themselves up as martyrs. 'Sainthood' has often been conferred based on horrific suffering or a martyr's death. Many religions, including Christianity, still imply that suffering is necessary to purify the soul and intentionally encourage rituals and practices that feature self-inflicted pain. But as our quantum paradigm demonstrates, it is impossible for suffering to purify something that didn't need to be purified in the first place? (Ironically, some of the same churches that encourage suffering in poor counties currently claim that God wants followers in industrial nations to be rich and comfortable.)

Gnostic Christians, who were also among the very earliest of Jesus' followers, saw his death in an entirely different light than orthodox Christians. When we understand that everything in existence *is* Divine Presence, we see the impossibility of offering anything to Source, including obedience, worship or sacrifice. Only in duality would it be possible to believe that the physical death of one being could atone for anyone else, let alone millions of others. Because Jesus' gnostic followers had experienced the Divine just as he had, they knew that each of us must be our own savior. For these followers, the death of a body known as Jesus meant nothing compared to the fact that he had awakened to Self and returned to Oneness. Sri H. W. L. Poonja, a 20th century sage better known as Papaji, explained, "You are the unchangeable awareness in which all activity takes place...Always rest in peace. You are eternal Being, unbounded and undivided. All is well...You are happiness, you are peace, you are freedom. Do not entertain any notions that you are in trouble. Be kind to yourself." Simply put, *suffering is another of our own creations*. As Hafiz asks, "O wondrous creatures, by what strange miracle do you so often not smile?...All your sufferings are from believing you know better than God. Such a special brand of arrogance as that always proves disastrous...There is a beautiful creature living in a hole you have dug."

When the Romans demanded that everyone burn incense to the emperor or die, many early Christians died rather than comply. The emperor was considered a God, and they reasoned that it was an act of worship that would anger their jealous God. Jesus' gnostic followers understood that the act was meaningless to Source because everything is Source, everything in existence is Divine. They burned the incense and lived. Knowing that the Divine is only interested in our awakening, not our obedience, worship, suffering, sacrifices or martyrdom eliminates the 'fear of God' so commonly associated with religion. (For a gnostic perspective on Jesus' life and death, we recommend our book, *The Beginning of Fearlessness: Quantum Prodigal Son.*)The writer of the gnostic *Gospel of Truth* promised, "Happy is the one who comes to [him/herself] and awakens." Rumi agreed saying, "What a joy to travel the way of the heart...Knock on the inner door. No other will take you to joy." Of course the reason for joy is the realization that this world is an illusion. It comes from knowing that we project our own virtual suffering and we are free to stop whenever we want. This is the key to spiritual fearlessness. Letting go of the little self and our attachment to this world while we are still projecting it allows us to live in it fearlessly. Literally returning to Self/Reality/Oneness ends fear forever.

One last thought on suffering: churches tell their congregations to pray for God's mercy, to beg God to end suffering. Ironically, this teaching cannot help but force a belief that God has the power to end suffering, but chooses not to. As a result, God is blamed for suffering because it was not prevented. Whenever tragedies occur, we hear religious leaders speak about "God's mysterious ways" or claim that God is allowing people to be killed in horrific accidents or die of excruciating diseases because it is God's desire to add a few more angels in heaven. They usually top off these explanations by assuring their listeners that "God is love." It's true that God is Love, but love could not abide any of these horrors. On the other hand, sages such as Jiddu Krishnamurti pointed out, "If...one discerns the true cause of suffering, that perception itself dissolves the very cause of suffering." Obviously few religions understand that we project our own suffering, but sages have been well aware of this truth. Suffering in this world feels very real, but if our world is not real, how could suffering be real?

Asceticism

Although it's usually not categorized as suffering, many are still turned off by the aesthetic life of denial that's often associated with spirituality. We see images of monks, gurus, priests and nuns who leave their families, given up their possessions, live in a monastery, a cave or other austere setting and beg for their food. Some religions demand that religious leaders give up intimate relationships or the possibility of having their own family. At the other extreme, it has become a part of New Age teachings that God not only wants you to enjoy abundance, but it's your duty to become wealthy. Again, each of these extremes in belief and behavior are the 'brainchild' of the little self. Both tie us to the body and keep our focus on obtaining blessings through action rather than a willingness to open ourselves to Reality. *It is not what we have or don't have that matters. It is not that we should own nothing, but that nothing should own us.* As the self gives way to Self, our preferences change, usually moving us toward simplicity. But simplicity has nothing to do with discomfort, nor does our choosing discomfort gain us any points with the Divine. Some might claim that we can concentrate more intensely on spiritual things if we avoid intimate relationships and family life; others feel certain that sexual and family relationships are a perfect place to grow. As we've said before, Source is interested in our return, not the way we live in this world. Just ask yourself if something you have, want or are doing is taking you closer to, or further from, love and oneness, and you will know what to do. Usually you will discover that it's your attitude toward the thing, not the thing itself, that is in need of change.

Earth School

Thinking of this world as a school means there also must be a cosmic teacher who assigns lessons and gives tests. There's no doubt that our world overflows with problems and difficult circumstances, and we can certainly learn from them if we choose to, but they're not sent to us by Source to either teach us or test us. As the 'stuff of the Divine' we share the One Mind. We have nothing to learn; much to remember, but nothing to learn. The One Mind has always been perfect, when we tune back into it, we will know. There was never a need to send us to the earth to learn, grow or evolve spiritually, especially not to a school that uses suffering to teach its lessons. We realize that 'earth school' is an extremely popular metaphor, but it's accurate only when we see ourselves as the school principal, the teacher *and* the student. What we experience and view as lessons or tests are either the result of random happenings, or they are something we inflicted on ourselves. When we're willing to examine our thoughts and experiences, we'll probably benefit. When we choose to ignore something of value, we might experience a consequence. Instead of a school, this world is our laboratory and we are the scientists conducting the experiment. Unfortunately, we also chose to be the lab rats!

Earlier we compared the pursuit of specialness to the lottery, where many lose so a few can win. In very real terms, this means that far more of us will 'suffer' in this world than not. If we're special in one area, that certainly doesn't preclude us from suffering in another. No one would bother to play the lottery if we took the odds seriously. To play we have to have at least some belief that we have a chance of winning even though the odds are heavily stacked against us. The same is true of this world. We must have been so convinced that we would be the 'lucky one' that we gave very little thought to the natural results of our dualistic arrangement before beginning our experiment. Regardless of the pain that duality dishes out, the amount of suffering we experience on the spiritual path is up to us. As Nisargadatta observed, "The mind misunderstands; misunderstanding is its very nature. Right understanding is the only remedy, whatever name you give it." When we cling to our misperceptions and feel like we're losing something important to us, letting go can feel difficult and even painful. When we see our misperceptions as valueless burdens we no longer need to carry, we can willingly let go of them and the journey is a pleasure. It's not uncommon to have some pet beliefs that are more difficult to let go of than others. We certainly found that to be the case, and caused ourselves some grief over them. But know that once you decide that you're ready to wake up, a few things might slow you down, but nothing can stand in your way.

Part Four

Waking in Reality

I. Beyond the Physical

The universe may have a purpose, but nothing we know suggests that, if so, this purpose has any similarity to ours—Bertrand Russell

Approximately 1,200 years ago, Shankara asked the question, "How can the physical eyes see anything but physical objects?" His question was rhetorical; he knew that the eyes, like all the body's senses, are limited. The body can tell us nothing more than what it perceives and perception is limited to the virtual reality of the material universe. Going past the senses, Shankara reconnected with the Divine and realized, "You are...the pure unchanging consciousness, which pervades everything...The Self is pure consciousness...You are the Self, the infinite Being...How can the mind of the enlightened think of anything other than Reality?" Although physicists have discovered the quantum world through the senses, their senses can never show them what it is. A handful of physicists have realized this and stepped beyond what the senses tell us. After taking that step, they give the same testimony as the sages. At that point they too have understood that the most exquisite deception employed by the little self is the illusory material universe of separate form.

As Shankara pointed out, the body and the little self are not the route that takes us to Reality, but getting past what the senses tell the brain, and what the brain wants to believe, is the greatest hurdle we each face. We've already talked a great deal about the way the brain functions, nonetheless, we may still believe in the brain's capacity for rational thought. But in light of our quantum reality, much of the brain's logic and ability to reason loses its value. The 19th century American satirist Ambrose Bierce was correct when he described the brain's logic as "The art of thinking and reasoning in strict accordance with the limitations and incapacities of human misunderstanding." The brain is a marvel of virtual reality, but its crucial limitation is its own function: believing the world is real. Author Frank Herbert observed, "Deep in the human unconscious is a pervasive need for a logical universe...the real universe is always one step beyond logic." Nisargadatta recognized the brain's limitation when he said that it, "...misunderstands; misunderstanding is its very nature. Right understanding is the only remedy, whatever name you give it." And as we've learned, corrected understanding comes to us by reconnecting with the One Mind. Plotinus understood that consciousness connects us to Source when he explained, "We do not pass through the material world; the material world passes through our eternal consciousness...Those who take the upward path divest themselves of all they have put on during their downward journey...We must then pass on upward, removing all that is other than God until...we behold the source of life, Consciousness, Being." Neither the brain nor the tiny portion of consciousness used by the little self can show us Reality. As a result, the little self sees the body, and especially the brain, as a perfect tool to keep us locked in virtual reality.

Getting past what the senses tell us is one of the greatest hurdles we each face. Why? We can learn intellectually that this world is a projection of consciousness and we can believe that this is true, but *the willingness to experience and live that truth is quite another thing.* Currently, spiritual interest is high, part of a cycle that has both ebbed and flowed throughout human history. It might seem logical that whenever spiritual interest rises, the little self would do something to divert our attention. It does offer endless distractions, but it also offers a more subtle substitute that appeals to the brain's logic and the body's desires. Rather than remove the desire to understand our existence, the false self convinces us our spiritual thirst can be quenched in this world, that meaning and purpose are found through the body rather than apart from it. This substitute works so well because it's based on our attraction to all things related to the body and our fear of the 'unknown' that lies beyond the body.

As you've already learned, religion serves the purpose of the false self by convincing us that we are unworthy of experiencing the Divine directly. But don't think that the little self has been put out of business by a shift from religion to 'spirituality.' The little self can just as easily use spiritual teachings and practices to keep the brain and body busily focused on this world.

II. The 'little self' and the Journey

Fear and Drama

You've probably read a scary book, attended a frightening movie, gone upside down on a roller coaster, skied down a mountain slope or perhaps even parachuted out of an airplane. Why? Since we live in a world that appears to provide many legitimate reasons to fear, the little self encourages us to feel comfortable with it by providing us with a level of fear and drama that's fairly safe. The fear excites us and when it's over, we feel that we've conquered it. We often get the same rush vicariously. Sensational news stories and 'reality' TV programs entertain and titillate us with lurid details of someone else's misery, while at the same time giving us the feeling that we're somehow special because we're safe and comfortable.

But what has our fascination with fear got to do with spirituality? To keep us distracted from Reality, the little self creates highly entertaining 'pseudo-spiritual' myths that predict both catastrophic and utopian outcomes for the future. Eight hundred years ago the Sufi sage Shams Tabrizi warned, "There are more fake guides, teachers in the world than stars." His words are still true and are worthy of keeping in mind whenever a new 'end of the world' prediction surfaces. Often these predictions are based on some 'spiritual revelation' or 'secret wisdom revealed to a special few.' Millions are kept on the edge of their seat chewing their fingernails, but when the catastrophe fails to materialize, it's quickly forgotten and another sensational prediction takes its place. On the other hand, we can get just as deeply involved in the belief that life on earth is about to take a dramatic change for the better, and it's our spiritual duty to play a part in that change by evolving into higher, more loving creatures. Currently there's a belief that positive energy has more power than negative energy. Eventually the world will be filled with enough positive energy to reach a tipping point where all negative energy disappears. The result of this spiritual/human evolution is supposed to leave the earth in a utopia of love and oneness. Before we give ourselves over to any of these scenarios constructed by the little self, let's let the quantum model shed some light on the subject.

For predictions to work, the future would have to be set in advance. This means that free will would give way to fate, destiny, predestination or determinism. Fatalism, destiny and predestination share the common belief that whatever happens has to happen, because nothing else can happen. These concepts are tied to the belief that there is a greater power in charge whose will cannot be altered or challenged. Since fatalism, destiny and predestination all render humans powerless and without choice; these belief systems encourage their followers to quietly accept their lot in life as inevitable. Determinism is slightly different; it's based on cause and effect and is therefore rooted in the denial of a universal Source. Determinism proposes that what's happening today is the inescapable result of things that happened in the past. Although determinists believe humans can play a role in creating a cause, they supposedly have no ability to alter the effect. Most proponents of determinism believe that cause and effect are so far removed from each other; we don't see the connection and falsely believe we have free will. When DNA was first discovered, some scientists claimed that it determined our choices. No doubt you've heard the claim that human behavior is 'hard wired' in the brain and we're little more than Neanderthals in dress clothes. Since then, research has discovered that genetics are better thought of as a 'probability' than a done deal. But Regardless of their differences, all these belief systems claim we're mere puppets who have no real say in our own thoughts and actions.

Although the quantum universe does exist within a set of 'operating parameters' that keep it intact, contained within those parameters is a sea of completely unpredictable potential. As the renowned physicist Werner Heisenberg pointed out, "The atoms...themselves are not real; they form a world of potentialities or possibilities rather than one thing of facts." As we've learned, it's the free will choices made by consciousness that act on energy/potential and bring the material world into existence. As a result, the only thing that remains unchangeable is the foundation of who and what Source is; everything else is subject to conscious choice. The only accurate 'predictions' that could be made are ones that foretell that change is inevitable and nearly anything is possible. Since the Divine is not interfering in our dream or giving secret messages to anyone, this also means that there are no groups or individuals who have special knowledge of the future. Of course some people are better at calculating what may be probable, but everything we see is subject to change up to the instant that consciousness makes a choice.

It's comforting to believe that positive thoughts and emotions have more power than negative ones. No doubt that's why many teachers claim that this is true, however, the belief doesn't stand up to the quantum litmus test or free will. In her book *The Intention Experiment,* Lynne McTaggart highlights numerous studies that demonstrate our thoughts do indeed have power and they can influence or affect others, but she points out, "A hateful intention is just as likely to cause harm as a loving intention is to heal." Positive and negative have equal weight or this world would not be an accurate description of our core intentions. It would also be impossible for free will to exist if one person's thoughts could override or overpower another person's thoughts. Free will would be impossible if my thoughts had the power to change you, and vice versa. Much as we don't like having someone hate us, we can't force a change in their feelings simply by loving them. We might influence them in a positive way so that *they* choose to change their thoughts about us, but our thoughts can't overpower theirs simply because they're positive. When gurus tell their followers that the thoughts of a few spiritual avatars have the power to negate the world's evil, their teaching is out of sync with quantum physics. And as far as any scientist has been able to discover, there are no overrides built into the system for goodness.

Research has shown that highly focused intention shared by a large number of people does appear to bring about the most powerful results, but those intentions can just as easily be negative as positive. Regardless, group intention still cannot force a change in someone else that they do not welcome. It does appear at times that negative thoughts have more power, but the real issue is how many people are thinking a thought. For example, as long as the majority of people live in fear, this will be a fear filled world. Let's keep in mind that our actions can be very different than our deep inner thoughts, and it's our thoughts that create, not our actions. This also explains why 'doing good' continually fails to solve the world's problems; although the effort appears to be sincere on the surface, core intentions for specialness and self-interest take precedence in our projections. It is true that if the vast majority switched from fear to love, the world could look very different. But still, that would not change its purpose as a laboratory to experience specialness. To escape that situation, we must return to the Self that is love, is peace, is Oneness and joy.

The world we see is a perfect mirror. It reflects an amalgamation of the core thoughts of our collective consciousness, no more, no less. What we want at our core is what is projected into the world. Although most of us believe that it is 'others' who are contributing to the problems in the world, few of us consider ourselves to be a 'negative influence.' As Voltaire put it, "No snowflake in an avalanche ever feels responsible." But as we explained in the last paragraph, what we project comes from our most deeply held desires, not from our surface thoughts, words or actions. The little self encourages us to believe we are what we say and do, not what we are thinking, but that is not the case. That's why Jesus posed the questions, "If you love those who love you, what reward have you? Do not even the tax collectors do the same? And if you salute only your brethren, what more are you doing than others?" (Matthew 5: 43-47) If we love those who love us but hate others, if we're still pursuing separation and our own specialness while we're doing 'good deeds,' we're sending that out into the world even if we never speak of it or acknowledge it to ourselves. Being a 'good person' that doesn't purposely harm others does not mean that we send only positive thoughts into the world. *Until we let go of everything that is not love, we'll project some level of desire for separation and specialness into the world.* Since that's the case, it should come as no surprise that our virtual reality includes the very best and the very worst. Although positive thoughts don't have more power than negative thoughts or vice versa, like pennies in a piggy bank, each thought does add up. If the ratio changed between the number of positive and negative thoughts that are projected, we would see a corresponding difference in the world.

The universe continues as it has for billions of years without needing even the slightest tune-up, but the world continues on its rollercoaster ride. Throughout history we've avoided Reality by titillating ourselves with predictions of a coming 'golden age' that's always right around the corner or a catastrophic event that's on the verge of destroying the earth and/or every living thing on it. As we all know, none of the predictions concerning a worldwide utopia have ever come true. And if we're realistic, we realize that as long as we project a dualistic system, it can never happen. However, predictions of global destruction could easily become self-fulfilling prophecy as we edge closer than ever to causing our own demise. Now instead of baseless predictions, scientists, economists and politicians are warning us that we're very near the tipping point of ecological, financial and social disaster. Their dire expectations based on hard evidence, can't easily be ignored.

Can we, or will we, destroy the earth? The entire point of our experiment is to prove that separation and specialness can outdo Oneness and equality. If an experiment is stopped before it reaches its conclusion, it proves nothing. Although Source knew that our experiment was doomed to fail before it began, the question would remain eternally unanswered unless our 'research' was allowed to run its complete course. Since we're experimenting both individually and collectively, any of us can come to the conclusion that separation isn't working and opt out at any time we wish. Baruch Spinoza made this clear when he said, "As stars high above earth, you are above everything distressing. But you must awaken to it. Wake up!" But collectively, the experiment will continue until either everyone 'wakes up' or it reaches its own catastrophic conclusion. But can this be considered a catasrophy?

Certainly if we bring about the destruction of the earth or life on it, we've proven beyond a shadow of a doubt that our way of doing things has failed. As grim as that sounds, we must remember that the failure of our experiment ends the little self and its projections, but can't harm the Self. And in fact, all that will end is an illusion, a projected thought no more real than a TV show or video game. As physicist Erwin Schrodinger put it, "We do not belong to this material world...We are not in it; we are outside. We are only spectators." The material world is a vibration that can be detected by the senses; not a reality that can actually be destroyed. At the theater, when the movie ends, the screen is still there; ready to receive another projected image. This has always been the view of sages who experienced the Divine. *The Tibetan Book of the Great Liberation* explained, "It is only because of deluded ideas, which you are free to accept or reject, that you wander in the world." Nisargadatta adds, "In consciousness the world arises; in consciousness the world appears and disappears."

We're not predicting one outcome or another since free will choice always has the power to change whatever direction the world is currently taking. However, each day you do have a choice between gently waking up on your own, or waiting until you're rudely shaken awake by the possible cataclysmic end of our experiment.

Eknath Easwaran advised, "There is an infinite changeless reality beneath the world of change...discover this reality experientially." But regardless of the method of your awakening, your safety is guaranteed. Since that's the case, getting wound up and distracted by man-made predictions and losing ourselves in the drama does nothing to help you on the spiritual path.

III. Doing or Being?

The little self uses the body to prove we're separate from everyone and everything else in the universe. This physical manifestation is so convincing, few of us see any reason to question it, even spiritual seekers. We might think that there is more to us than what we see, but we still accept the body as our reality. When we understand that the body is a puppet or avatar that we project, it becomes easier to let go of our attachment to the body and personality that's been constructed for it. The little self wants us to think the body is the path to salvation through 'doing.' Taking action, making things happen, doing rather than being, keeps us focused on the physical and away from the inner path, which requires no physical action. The human animal was a perfect tool for the little self to coopt, since, like all other animals, doing is its nature. In contrast, the Self, as pure consciousness, creates through thought, and cannot be said to *do* anything. Simply said, the body/little self is a 'doing,' the Self a 'being.' For that reason, transcending the physical comes about through the alteration of the level of consciousness that's responsible for our thoughts, not the things we do. And of course, choosing to open the door to the higher consciousness of Self, has nothing to do with physical activity. But to keep us in this world, the little self does its best to keep us away from 'going inward' by convincing us that spirituality requires 'going outward.'

Works

Many religions, Christian and otherwise, support the principle that "faith without works is dead." This concept does encourage many good works, but it denies the fact that we are already recipients of the free gift of Divine grace. Many believe that grace equates to God forgiving us even though we are unworthy sinners. From a quantum perspective, grace means that we always have been and always will be worthy because we are one with the Divine. Hafiz made this clear when he said, "God wants nothing in return for your existence. What madness to ever think you owe Him/Her/It anything...Mercy and Grace are just there; they are attributes of light, they want nothing but to be…This is the time for you to compute the impossibility that there is anything but Grace." Grace is a given, an attribute of the Love that Source is.

Kind deeds that flow from compassion are always positive and worthwhile. Whatever deeds a loving heart is moved to perform should be undertaken. Nevertheless, if works alone could change the world, it would have been in far better shape than it is, long ago. Immeasurable quantities of time and money have been thrown at the world's ills, and still they persist and multiply. Why haven't good works solved world problems? As we said earlier, the catch is our core thoughts. Because it can be uncomfortable to confront our deepest thoughts, we may have hidden them from our own awareness. Nonetheless, the quantum world responds to them and that is what we project. In a conscious universe where thought, not actions, create, it's also the motivation behind the act that has the power. When 'good works' are born from selfish motives they can help the recipient, but they will fail to change the doer. When we support some causes but reject others because we're judging the recipients, our efforts can't result in permanent change. This is true of our efforts toward peace as well.

The Dali Lama explains, "If we ourselves remain angry and then sing world peace, it has little meaning. First, our individual self must learn peace." Jiddu Krishnamurti described the importance of thought when he observed, "I am the world and the world is me. If I am corrupt I will create a world which is corrupt, or I will sustain a society which is corrupt...What you are the world is. And without your transformation, there can be no transformation of the world...It is love alone that leads to right action. What brings order in the world is to love and let love do what it will." Aldous Huxley confirmed this truth when he said, "I wanted to change the world. But I have found that the only thing one can be sure of changing is oneself." Since this world is the reflection of what we are at our core, *being* Love is the greatest gift you can give.

Practices

Doing is not confined to good works. Anything that requires the actions of the body falls into the category of doing. Waking up is the realization that we are not the body, but doing in any form can easily reinforce the belief that we are. No amount of 'doing,' no matter how spiritual it seems, can take you back to Self. The key here is whether or not the 'doing' is something that *supports* you as you let go of attachment to the body or reinforces it. If a practice becomes an end in itself, it can easily take you deeper into the little self. Going on a pilgrimage, walking a labyrinth, creating rituals or 'being in the now' all appear to be harmless, but if they're not helping you let go of the thinking that keeps the little self in charge, they're taking you on an unnecessary detour. As the 20th century priest and psychotherapist Anthony de Mello wisely pointed out, "The donkey that brings you to the door is not the means by which you enter the house." For many, the donkey becomes so important, the house is forgotten.

Many practices are based on the false premise that we need to either prepare or purify ourselves before we can approach the Divine. Only after years, or even an entire lifetime, of rigidly disciplining the body and mind is a practitioner considered worthy. But sages who have had a direct, personal experience of the Divine say that preparation is unnecessary; the Divine accepts us exactly as we are. As Hafiz promised, "There is nothing you have ever done that is not innocent and will in any way be judged as wrong by anyone of true wisdom." Yoga and meditation are both popular practices that focus on controlling the mind by controlling the body. There are certainly no rules against using these methods if they help you, but there are also no valid rules saying you must. Yes, many teachers claim yoga or meditation are absolutely necessary, but far too many have experienced the Divine without engaging in these practices for that to be true.

Since yoga and meditation were introduced to the West, they've both undergone rapid change. Some alterations have been so drastic; the practice has lost its original meaning and is now better understood as a relaxation technique, stress reducer or exercise regimen. Around the 2nd century BC, Patanjali compiled a group of sayings known as the *Yoga Sutras*. Yoga is a Sanskrit word that roughly translates as union, yoke, join or unite. According to Patanjali, yoga brings the true Self (Atman) to a state of union with Ultimate Reality (Brahman) by restoring our awareness that Atman and Brahman are one. Patanjali described five forms of yoga: karma, jnana, bhakti, raja and hatha. In the West, we're most familiar with hatha (reaching the Self by controlling the body) and raja (the path of the mind, primarily using meditation). As we've said, many gurus insist that meditation or yoga is absolutely essential, but the ancient *Ashtavakra Gita* claims just the opposite saying, "The practice of meditation keeps one in bondage. The body is strained by practices. The mind numbs with meditation. Detached from all this, I live as I am...The liberated one does not exert effort to meditate or act. Action and meditation just happen." Simply put, action or meditation may or may not arise naturally as we return to Self, but they are *not* necessary.

For some, these practices are not only a distraction, they can be downright discouraging. A recent study on the benefits of meditation reported in the Journal of Consulting and Clinical Psychology, reported that 54% of the participants felt more anxious when meditating. This can especially be true for anyone who is hyperactive or has ADD. Perfecting a practice can also demand all our attention. Striving for perfection is extremely appealing to the little self, who would like to convince us it's more important to be able to sit perfectly still for 12 hours than experience the Divine while we're out taking a walk. Since the little self thrives on specialness, it will happily push us to compete with others who are also trying to perfect their skills. If you feel that a practice is helping you let go of misperception, use it. If a practice distracts you from letting go, leaves you with a feeling of superiority, competition, frustration, unworthiness or even failure, it's serving the little self.

Knowledge and Intellect

Viewing spirituality as an intellectual pursuit is also a form of doing. Studying, reciting and interpreting holy books and the endless squabbles over texts that take place both within and between religions can lead to exceedingly long detours. As the *Bible* book of Ecclesiastes points out, "The sayings of the wise are like goads, and like nails firmly fixed." However, the writer also warns, "My son, beware...Of the making of many books there is no end, and much study is a weariness of the flesh." (Ecclesiastes 12:11, 12) As you can probably tell from the numerous quotes we're using, we have an extensive collection of religious and spiritual writings and we've spent a great deal of time reading them and contemplating what they say. However, it would all be a waste of time without the direct experience that is our foundation; the words are just a means of hinting at the truth discovered by experience. As the gnostic *Gospel of Truth* explains, "Those who have realized gnosis know the source and the destination. They have set themselves free by waking up from the dream...and have become themselves again." No one can convey the essence of Divine experience with words, but words can demonstrate that the experience is possible.

It's important to remember that no matter how wise a 'sacred writing' appears to be, there are no books, no matter how old or revered, that contains a message sent from the Divine. As we've learned, Source is not going to meddle with our free will or play favorites by giving an exclusive message to a 'special person' or a 'favored group.' Instead of seeing these so-called 'sacred books' as the direct communication of God, realize they may either be the words of seekers who experienced the Divine directly, or the interpretation of those words made by their students and followers. Remember these words have usually been altered numerous times by those who didn't understand what the sage was teaching, or have been twisted by those who wanted to use them to serve their own purpose. Once you've experienced the Divine, you will be ready take the advice of the *Srimad*

Bhagavatam, an ancient Hindu text written about 1000 CE, "Like the bee gathering honey from different flowers, the wise man accepts the essence of different scriptures." In this case, the operative word is 'essence.' Instead of being a 'literalist' who takes each word verbatim, you'll be able to discern the truth that goes beyond the words.

Another intellectual 'doing' the little self enjoys is quickly hopping from one new spiritual idea to another. The brain easily mistakes collecting tidbits of information with progress; the more trivia it amasses the more progress it thinks it's made. The little self can easily convince us that we must continue to investigate everything that comes along, but it's actually just keeping us on the surface. Throughout history there have been religious and spiritual fads, and in the internet age they spread like wildfire and disappear just as quickly. We can easily get caught up in the excitement of each new idea, guru, teaching or practice as we do our best to keep up. The little self encourages the idea that someone has uncovered a hidden secret that will miraculously awaken us. Of course this smacks of specialness, but that's one of the draws. There is *no* magical spiritual pill, and certainly nothing that can do the work for us.

Rumi cautioned us about "...wander[ing] from room to room hunting for the diamond necklace that's already around your neck." The activity of the brain and body easily convinces us we're very busy spiritually, but it's more like exhausting ourselves on a treadmill that goes nowhere. In our work we've come into contact with many spiritual seekers who have been 'looking' outside themselves for 10, 20 even 30 years or more. As a result, they're always afraid they'll miss the 'real' answer if they don't keep looking. They bounce from one thing to another so quickly, they remain spiritual infants. Their constant pursuit has kept them from realizing they need to quit skimming the surface and dive deep inside. Rumi reminds us, "You're looking for the Pearl? Plunge, now, to the sea's bottom. What's on the shore is only foam." Notice that the wisdom of ancient sages still rings true today:

Those who are interested in the design of the jars are not thirsty—Saadi (medieval Persian mystic poet 1184-1291)

These spiritual window-shoppers who idly ask "How much is that?" They handle a hundred items and put them down—Rumi

There will always be lots of gab along the shore from those who are drawn to God but have yet to get bare assed and go in—Tukaram (Hindu mystic poet 1608-1650, thought to be the first to deny the caste system)

Few are those who reach the other shore; most people keep running up and down this shore—*Dhammapada* (teachings of Buddha)

Few cross the river of time and reach non-being. Most run on this side—Horace (Roman poet philosopher, 65BC-8BC)

Nisargadatta sums up the reasons why hopping from one guru, fad or practice to another reinforces our attachment to the little self, "Running after saints is merely another game to play. Remember yourself instead and watch your daily life relentlessly. Be earnest. Your own Self is your ultimate teacher. The outer teacher is merely a milestone. It is only your inner teacher that will walk with you to the goal, for [Self] is the goal." Our purpose in writing is not to have you think of us as teachers. Rather, our aim is that our books will act as diving boards you can spring from to enter your own deep and fulfilling experiential adventure.

IV. Spiritual Evolution?

Currently, many teachers claim that we're entering a new age of understanding. They believe God is in the process of evolving human consciousness to a higher level so we'll be more in tune with the Divine. This view implies that we were not previously given a level of consciousness that would allow us the spiritual understanding needed to escape suffering. Of course this idea is once again based on a God who manipulates this world and controls our destiny. To accept this view, we would have to let go of the quantum reality of One Mind and the concept of free will. While we feel certain that these teachers have only the best intentions in mind, they've forgotten what countless spiritual masters have demonstrated: *the Self already shares the One Mind of God, and that Mind does not need to evolve.* While it is obvious that levels of awareness exist in the *material* portion of the universe, consciousness either is or it isn't. And we cannot forget the fact that *the material portion of the universe is not the seat of consciousness.* Raising the levels of human consciousness does not make sense because consciousness does not reside in the human brain. Although some scientists insist consciousness evolved from matter, quantum physics tells us that *consciousness is the universal ground that permeates and supports everything in existence* and is the essence of even the tiniest subatomic particle.

When we decided to experience separation and specialness, it was necessary to close the door of access to the fullness of the Divine Mind. As a result, we are not aware of what full consciousness could offer us, but that alteration did not negate or diminish the existence of the One Mind or its infinite knowing. We chose not to partake of that fullness, but our choice does not mean the One Mind needs to evolve. Einstein rightly realized, "Something deeply hidden has to be behind things...The superior reasoning power...revealed in the incomprehensible universe, forms my idea of God." Scientist and inventor Nikola Tesla felt the same way saying, "My brain is only a receiver, in the Universe there is a core from which we obtain knowledge, strength and inspiration." It's true that we do need to reconnect with the fullness of the One Mind, but the Divine will never override our choices and open that connection against our will. However, if we think consciousness needs to evolve, we're actually saying that the One Mind is in a growing and learning process and is using the evolution of the material universe as a tool of self-education. If that were true, we would have to conclude that God was going through an endless series of 'growing pains' that we endure. Regardless of these suppositions, when we reopen the door to the One Mind, we realize 'evolution' could never heal the essential flaw of separation that was brought about by dualistic thought.

Some teachers say that God is orchestrating an evolutionary change so that a 'higher human form' will become capable of creating a utopian world. If that were true, anyone with compassion for the suffering and misery that has been endured for centuries by 'un-evolved' humans would be justified in demanding, "What the hell took you so long?" And why would evolution even be necessary? Couldn't an all-powerful being just will it to be so? Most of these teachers claim that while God is evolving humans into something more loving/spiritual than humanity has ever been, it is also up to humans to willingly make the changes themselves. Both can't be true, but the idea becomes even more complicated by the belief that this evolution focuses on the outer rather than the inner. How is this evolutionary change to love and oneness supposed to come about?

The main argument used to support this theory goes like this: if you do the work you love, have the things you desire, live in comfort, perfect your body, and become more loving, the changes you make will have a domino effect that will eventually heal the world and bring about a utopia. It's no real surprise that the theory of spiritual evolution has great appeal for the Western mind that has been taught to equate an increase in self-satisfaction and personal comfort with progress.

This philosophy could have some merit if we actually used what we acquired for good and become the love and peace the world so desperately needs. But for the most part, this view is a rationalization that supports narcissistic self-indulgence and keeps its adherents locked into 'doing' and the pursuit of separation and specialness. Sadly, this way of thinking has permeated our culture so deeply; we think that almost anything we do to improve our own life, no matter how hedonistic and shallow, can be considered spiritual. Is it any wonder that businesses have picked up on this trend and now advertise spa treatments, plastic surgery, clothes, cars etc., etc., as something that can nurture your spirit? No doubt you've seen or heard advertising for products as mundane as body lotion or soap that claim to nurture 'mind, body and *spirit.*' Needless to say, nothing material has anything to offer the Self. *Few things keep us off track as easily as confusing the physical and the spiritual, the outer with the inner.* This is nothing new. Teresa of Avila, a 16th century Carmelite nun and mystic aptly observed, "Our body has this defect that the more it is provided care and comforts, the more needs and desires it finds."

As the quantum model has taught us, we are not the body and in Reality, the body doesn't exist as anything more than vibrational energy. As Plotinus pointed out long ago, "Those who identify the body with Real Being are like dreamers who mistake figments of sleeping vision for Reality...*True waking is not of the body, but from the body.* Anything else is just a passage from sleep to sleep." [italics ours] Plotinus was not alone in his understanding:

The body is a painted image subject to disease, decay and death, activated by thoughts that come and go...Remembering that this body is like froth, of the nature of a mirage, break the flower-tipped arrows of Mara (illusion)—Buddha

There is not the least hope of liberation until [we] stop mistakenly identifying [ourselves] with the body—Shankara

Don't worry so much about your body...Leave it as is. Be more deeply courageous. Change your soul—Attar of Nishapur, 12th century Persian mystic poet, author of *Conference of the Birds*

Break free from the chains you have forged about yourself. You will be free when you are free of clay—Hakim Sanai, 11th century Persian mystic poet, author of *The Walled Garden of Truth*

The ultimate value of the body is that it serves to discover the cosmic body, which is the universe in its entirety—Nisargadatta

Nisargadatta was spot on when he said, "To see the universe as it is, you must step beyond the net. It is not hard to do, for the net is full of holes." If we think of the material portion of the universe as the netting, we realize the net is insignificant compared to the holes. As far as researchers can tell, the universe is made up of 70% dark energy, 25% dark matter, which appears to be a structure for matter, and only 5% matter. Dark matter and energy are not literally dark, but got that label because we currently understand so little about them. The world that's so important to us is an infinitesimally small portion of All That Is, a speck floating in the sea of Reality. As we've learned, Source is not involved in this world and is not evolving us or though us, and is certainly not dependent on us to improve the earth. Rather, these concepts serve the little self since they keep us distracted and focused on the illusion rather than the Reality.

However, since everything that exists, including the illusion we project, is the 'stuff' of God, it deserves to be cared for in a respectful way. Just as there is no benefit in perfecting the body or the world, there's also no point in rejecting the body, renouncing the world or living in a way that purposely harms either one. We can strike a realistic balance when we understand the difference in value between fixing up the illusion compared to transcending it. Of course realizing who and what you actually are makes that evaluation much easier. Jesus recognized that while he and his followers were "in the world" they could be mindful that they were not "of the world." (John 17:16, 18:36) Jesus explained how his followers could maintain their priorities when he told them to "seek first the kingdom...and all these other things will be added to you." Keep in mind that the things that Jesus said would be "added to you" were the necessities of life, not the luxuries. (Matthew 6:33) Jesus' Jewish followers already believed that God would conquer their enemies (the Romans) and establish a literal kingdom on earth. It's not surprising that they misinterpreted his words and thought he was talking about a physical rather than spiritual kingdom. In the gnostic *Gospel of Thomas*, they anxiously asked, "When will...the new world come?" Instead of encouraging them to look forward to a government to cure their problems, Jesus answered, "What you look forward to has already come, but you do not recognize it...It will not come by waiting for it. It will not be a matter of saying 'Here it is' or 'There it is' Rather, the kingdom of the Father is spread out upon the earth, and men do not see it."

Although Jesus made it clear that his followers should not expect a physical kingdom, many of them were so sure a messiah was coming to save them they interpreted his words through their own longings to improve their life in this world. They decided Jesus was the messiah they hoped for but when he was killed by the Romans, they decided that he would return within their lifetime to set up God's government.

When that didn't happen, they pushed the 'promise' of a theocracy so far into the future, apocalyptic Christians are still waiting for it. They may not call it 'God's kingdom,' but the teachings of many New Age gurus parallel this ancient desire for a utopian theocracy. Jesus was only one of countless sages who have clearly told their followers, "...the kingdom is inside you and it is outside you. When you come to know yourselves, then you will be known, and you will realize that you are the sons of the living Father." (*Gospel of Thomas*) In the gnostic *Dialogue of the Savior*, Jesus again pointed out that the kingdom was not literal when he said, "Every one [of you] who has known himself has seen it." These words take us right back to the truth that spiritual awakening has nothing to do with focusing on our projected illusion, improving it or renouncing it; i*t has everything to do with waking up from it.* Direct experience taught Carl Jung, "The way is within us, but not in gods, nor in teachings, nor in laws. Within us is the way, the truth, and the life."

V. Intention

Since the non-scientific community got wind of the power of quantum consciousness, intention, visualizing, manifesting and the 'law of attraction,' have become extremely popular.

Spiritual sages have always realized that consciousness is the foundation of the universe, and the tool of creation, but currently consciousness is being touted as a magical genie that's waiting to obey our every command and transform our desires into material reality. On the surface it sounds wonderful; just choose what you want, focus your attention, and the world is yours! We know that consciousness does act on energy to project material form, but is the 'law of attraction' an accurate depiction of consciousness at work? Personally, we've both had and seen some remarkable results that have been attributed to intention, but we've also seen it fail miserably and cause a great deal of pain. If the connection between consciousness and physical manifestation is as unbreakable as proponents of manifesting claim it is, why does intention appear to be both limitless and limited? 'Law of attraction' gurus generally place the blame on the wavering focus of the 'intender' or claim that a surface intention is in conflict with a more powerful hidden intention the intender is not aware of. But is there more to it than that? Yes, there's a great deal more.

Before a swimmer dives into unknown waters, they check to see what's hidden beneath the surface, and we need to do the same where intention is concerned. Although manifesting originated from the fact that conscious thought causes energy to appears as material form, the idea that we can use this connection to fulfill all our own desires has several flaws. First of all, it's imperative to realize this world is the result of choices made by 'group consciousness.' As Nisargadatta so wisely said, "Everything is caused by innumerable factors of which your personal endeavor is but one." As your intentions go out into the universe, so do millions of others. Since this world is a game of specialness, we would like to believe that our thoughts are more powerful than someone else's, which would give our choices precedence, but this is not the case. If that were true, two basketball teams that are contending for a championship would do better if they quit practicing and manifested instead. Then the more dedicated 'manifesters' would win, not the better team.

In the book *Quantum Enigma: Physics Encounters Consciousness* by Bruce Rosenblum and Fred Kuttner, the authors explain, "If someone looked in a particular spot and happened to see the atom there, that look 'collapsed' the spread-out waviness of the atom to be wholly at that particular spot. The atom would then be at that spot for everyone." Simply put, it's the first thought that affects energy, and once a thought has 'congealed' into material form, it is set in that form. This principle means that the first thought that brought a pomegranate into existence in this world, set that form and it will remain the same for everyone as long as the thought of pomegranates exist. The person who designs or builds a chair or table or bakes a cake sets that form, and we all see it the same way. If that weren't the case, this world would be in complete chaos, and it would be impossible for us to navigate our way through it or to conduct our experiment. But what does this mean for us in the big picture?

The primary thought that caused us to project our dreams onto the world of form was our desire for separation and specialness. As soon as we began using the human animal as an avatar, the material world began reflecting that thought above all others, and it will continue to do so as long as we keep that thought. And as we've discovered, that means we will live in a system very similar to the lottery where many have little so a few can have much. We can use intention to influence what happens to us to some extent, but only as far as the thoughts we project either mesh with prevailing thought patterns or at least aren't obscured by them. But let's face it, this sort of manifesting is nearly always directed at things that promote our own specialness. And, since we can't see around corners, the thing that we try to manifest today could be the very thing that causes us trouble and sorrow a year from now.

Intention is our tool of creation at the quantum level, but it is only when we return to Self that we are able to wield it at full power. As long as we continue to project this world our intentions do have power, but in a relative sense. For instance, our intentions can have a positive effect on our own health or help us accomplish something, but they can't force something to happen, or make someone love us or even like us. Often when we have an intention, we make changes that fit the intention and it may be that those changes assist us in bringing about what we desire. For instance, no matter how strongly we intend better health, we won't have it if we just sit in front of the TV eating chips and drinking beer. Simply put, using intention to imagine and create a perfect life for yourself is an iffy proposition at best, and certainly one that will take your focus away from spiritual awareness. However, intention, coupled with the willingness to let go of conditioning and allow yourself to be taught by Source, can be an extremely powerful tool. It's true the intentions of the little self are opposed to oneness, but our spiritual intention has the power of Self that outstrips anything the puny little self can come up with. Intention, which is a potent form of focus, can certainly help keep you on track. Still, intention alone will not miraculously awaken you.

There are other aspects of the quantum world besides intention that are at work in our virtual reality. Each body is surrounded by an electromagnetic field that resonates and interacts with other fields. University of Arizona researcher Gary Schwartz explains, "We are, so to speak, swimming electromagnetically in one another's...fields." Research is demonstrating that the heart is the strongest field generator, and it resonates with our core intentions. When we are love, our field resonates with that love, and it can affect others who come within that field *if they choose to partake*. Negative thoughts and intentions also resonate through our fields, whether we're aware of what they are or not. If you've felt uncomfortable around someone, but saw no reason why you should feel that way, it was probably something that you discerned on a subtle level when your electromagnetic fields came into contact. Even though we can 'feel' other fields, none of them can alter our field unless our own consciousness allows it.

Regardless of what gurus of manifesting claim, our consciousness did project, and does sustain, this dualistic system. At the very least, this should impress on you how powerful consciousness is, and how necessary it is to discern which thoughts are valuable and which are valueless, which you want to encourage and which you want to discard. Since intention is more about quantum physics than magical thinking; if you want to know more about the subject, we suggest that you get your information from balanced sources backed by scientific research.

VI. Your Own Path

Since religion offers a 'one size fits all' solution to humanity's problems, many assume that spirituality should work the same way. After all, it seems like the journey would be simplified if there was 'one true path' everyone could follow, one set of directions that would unerringly bring us all swiftly to the goal. It takes only a few minutes searching the internet to find many teachers who claim they've discovered just such a method, program, system or secret that, if followed carefully (and often at great expense), can lead everyone to enlightenment. However, those claims are as reasonable as expecting that a person in New York and a person in Los Angeles can each get to Chicago by following the same itinerary. It is true we're all returning to Source. As *The Kybalion* points out, "All are on the Path whose end is the All," but we've each followed a distinctly different route that led us to our current location. Since every path has a unique starting point, we must each discover our own way back to Source by walking on it. This road is not something concrete that's laid out in front of us, but something free flowing and organic that takes the shape that's needed as it's needed. Since there can be no specific itinerary that works for everyone, it is entirely possible to 'wake up' and gain mastery over the little self in complete isolation. Nonetheless, most of us appreciate getting a few tips from those who have successfully gone before us. But don't confuse those tips with rules or specific directions.

Religion has many adherents in part because the clergy class claims to be more qualified and able to work out 'salvation' for us. For the same reason, many are drawn to gurus and teachers because they see them as a ship they can ride on. But as the Buddha reminds us, "It is you who must make the effort, masters only point the way." And Lao Tzu accurately pointed out, "The Tao (way) that can be spoken of is not the Tao...You can't know it, but you can be it." Clearly, *reconnection with the Self is, and always has been, a 'have-it-yourself' experience*. Still, masters have written and taught so others will know that transcendence is possible. No master was born that way, none of them were sent from heaven or given special powers or told secrets that are unavailable to the rest of us. What they each want you to know is that you can accomplish exactly what they accomplished. If they could transcend this world, so can you. As Rumi so lovingly pointed out, "Ours is not a caravan of despair. Come, even if you have broken your vows a hundred times. Come, come again, come."

Most of us use the word 'perfect' to describe something flawless, like a diamond. From that perspective, a spiritual master is not perfect. Instead of being without defect or blemish, masters can be called such because they have mastered the little self. He or she has returned to Self by diminishing the little self to the point that it rarely surfaces. On the very few occasions when it does, they quickly recognize what's happening and let go of whatever thought supported its return. In a story told about Buddha, an angry man spit in Buddha's face while he was speaking with a group of students. Buddha listened politely while the man vented his rage. When he finished, Buddha thanked him for coming and invited him to come back whenever he felt like spitting. Why? Buddha explained the event was an opportunity to gauge whether self or Self was in charge. He realized that although he lived from Self, the little self could try to resurface as long as he continued to project a body.

As we've said so often, whether a spiritual master is involved in intellectual pursuits or not, their words are always based on their own direct and continued experience of Ultimate Reality. If we think of a spiritual master as someone who can master our little self for us, or even direct our steps, we'll be sadly disappointed. Not only do they understand that would be impossible, more importantly, they have no desire to do so. However, many who claim to be teachers but lack direct experience are eager to convince others that they can shape their path for them. Such a 'teacher' continues to operate from the little self. Although they may have gathered a great deal of information and have honed the ability to retell it; *they **know about** the Divine, but they do not **know** the Divine through personal experience.*

The 10th century Zen master, Yun-men succinctly explained to his students the difference between teachers and masters when he said, "No one can do it for you. Every one of you should work toward self-realization. Why should you preserve my speech and tie up your own tongues?" He also warned, "When you hear that some great master has appeared in the world to liberate all beings, you'll immediately clap your hands over your ears. As long as you aren't your own master you may think you have gained something from what you hear, but it is secondhand merchandise, and not yours. Those who really have it live like ordinary men. The master can only bear testimony. If you have gained something within, he can't hide it from you, if you haven't gained anything, he can't find it for you."

An ancient Tibetan proverb cautions, "A guru is like a fire. If you get too close, you get burned.

If you stay too far away, you don't get enough heat. A sensible moderation is recommended." Sadly, many make an idol of a teacher or master and come to see emulating their outward behavior or memorizing their sayings as the goal. No matter how hard we might try, *it's impossible to produce a cause by repeating the effect.* Mimicking outward behavior can never create inner changes. As Shankara pointed out, "Who has overcome the world? He who has conquered his own mind." *Believing in words or in people does nothing for us, taking self-responsibility and experiencing the Divine for ourselves does.* We don't want you to profess your belief in any of the words in our books. If that's the case, we've failed you. Our desire is that the words we've written and quoted will open a door of possibility that supports and encourages your own personal inquiry and experiential journey. As Rumi said, "I stopped asking the books and the stars. I started listening to the teaching of my Self."

Amazement

Addiction to the "spiritual experience" can be one of the most covert and subtle challenges for the spiritual seeker. Scientists have discovered that practices such as intense meditation not only alters brain waves, but it can also cause the pineal gland to produce a psychedelic drug known as DMT (dimethyltryptamine). This combination has the ability to produce visions that can be overwhelming, blissful and highly addictive. Meditators can get caught up in the drama and feel compelled to repeat the experience again and again. When this happens, the original goal of waking to Self is forgotten and the means becomes the end. This is especially true when the little self interprets the vision as 'specialness.' Some consider the pineal gland and the chemicals it produces to be the body's connection to God, but it's important to note that given the right conditions, anyone, whether they are a spiritual seeker or not, is capable of producing these visions. These alterations in brain waves and chemistry have nothing to do with the modification in consciousness that takes place when we reconnect with Self and the One Mind. Still, when the seeker reconnects with Self, they can experience the Divine in ways that are beyond explanation.

Jesus was talking about reawakening to Self when he said, "I shall give you what no eye has seen, what no ear has heard, what no hand has touched, what has not arisen in the human heart...Let one who seeks not stop seeking until one finds. When one finds, one will be astonished." (*Gospel of Thomas*) In some translations, the word 'troubled' is used instead of 'astonished,' and either word can fit the experience. When Jesus said, "I shall give you..." he was not planting these experiences in the consciousness of his follower. Rather, he was giving them living proof that a direct experience of the Divine was not only possible, it was available to all who were willing. When he told them to "keep seeking until one finds," he was reminding them that this was not something he could do for them."

The experiences Jesus spoke of cannot be called up at will; they happen spontaneously as a result of opening to higher consciousness and do not require meditation. Going from the consciousness of the little self to the Self could be equated to the changes that take place as you travel from sea level to a high mountain altitude. At first the heights may make you dizzy, but after you've been there for a while, you become acclimated. As your level of consciousness changes, these transformations *can* produce some spectacular experiences, but some go through the transition without any of the fireworks. Enjoy the experiences if you have them, but remember they aren't necessary, don't make you special and are something to pass through, not cling to. Regardless, once you reconnect with the One Mind, you will no longer see things as you did before. As Nisargadatta so aptly put it, "The search for Reality is the most dangerous of all undertakings, for it destroys the world in which you live." Spiritual awakening destroys the veil of illusion while opening our spiritual vision to the infinite. Although this transition can be both astonishing and troubling, no one who has made this transition has ever done so with regret.

The moment we think mystical experiences make us special, we put ourselves right back in the hands of the little self. Think of them as breathtaking panoramas along the highway. We may ooh and ah over them, and gain something from seeing them, but they're not the destination. As a consequence of altered consciousness, some find that they suddenly have what could be labeled 'psychic abilities.' In light of quantum research, it may be more accurate to say that this knowing is the result of fuller access to the One Mind. Earlier we discussed the vast amount of 'dark energy' in the universe. Physicists also call this quantum sea 'the zero point field.' This strange name came about because scientists wondered if space was a vacuum where nothing existed. As a test, they removed all matter from a defined area and then lowered the temperature to absolute zero (-273.15° centigrade). Theoretically, substances should possess no thermal energy at this temperature, but instead they found that the energy in a single cubic meter of space would be enough to boil all the oceans of the world. But there was something more. The zero point field also acts as an enormous repository of memory. The brain has been shown to act as a sophisticated retrieval system for some of this stored information, but a higher level of consciousness allows us to access much more of it. Some feel the ancient 'Akashic records' refer to this information repository. The word Akashic is translated 'sky' or 'space.' This celestial 'library' was said to contain a record of all human experience as well as the history of the cosmos, so it's no surprise that connection has been made.

When we understand so-called 'psychic abilities' from the quantum perspective, we can see that it's not a special ability, but a function of consciousness that. Although they may not have known what it was, some have always found that they can tune into the zero point field more readily than others. Hypnotism may also let down the brain's defenses so we can more easily access the zero point field. If we see psychic ability as another aspect of the transition from self to Self it won't create a distraction. However, if we allow ourselves to be overwhelmed by it and focus on it, it can take us on a detour that can keep us busy for this, or even several, lifetimes. As we continue to let more and more of the little self dissolve into Self, it's helpful to remember the words of Plotinus, "Those who take the upward path divest themselves of all they have put on during their downward journey. We must then pass on upward, removing all that is other than God until...we behold the source of life, Consciousness, Being."

Emotional Issues

In this world, some revere the intellect and allow logic to rule their lives while others are led by their emotions. Either extreme causes problems. We've all seen people who are stuck in either logic or emotion, or swing violently back and forth between the two. Most of us try to strike a balance between logic and emotion, but that can be as difficult as trying to keep oil and water mixed. When we're ruled by logic alone, everything becomes a balance sheet, including our relationships. Logic figures out the cost/benefit ratio of all experiences and holds back whenever it thinks something might be a bad investment. On the other hand, emotion is an energy that moves through the body and sets off complex physiological (chemical) reactions that the body can easily become addicted to. Surprisingly, raw emotion starts in the brain with thoughts that are allowed to run wild. These thoughts trigger chemical reactions, drawing the body into the process. We may find it odd that a person continues to worry when they have no apparent reason, but they may well have worried enough to set up an addiction to the chemicals their body creates as a response to worry. When we let our emotions take charge, we're erratic, irrational and perceive everything through the haze of feelings. Emotion jumps headlong into negative or dangerous situations without counting the cost.

Between logic and emotion, we've come up with some very strange concepts of what love actually is. We divide 'love' into separate categories, give them labels such as *eros*, *agape*, *filial* etc., and attach expectations and conditions to each type. The logical brain sees love as a cold, hard give and take proposition, while emotional love is so mired in gooey sentimentality and romance; it flips back and forth between agony and ecstasy. Love ruled by either logic or emotion has conditions and can quickly become hate. And let's not forget the role the body plays, interpreting love as lust. The brain sees love as something that must be controlled; the emotions and the body lose control. Neither of these extremes is satisfying, and neither are accurate depictions of love. By experiencing Divine love, the sage realizes there is a third option, something scientists are now identifying as 'emotional intelligence.'

The Western world has long claimed that the brain is the center of intelligence that rules the body, including the heart. But neuroscientists have discovered the heart begins to form before the brain, and it beats on its own without the brain's direction. Most astonishing, *the heart has its own intelligence*. The forty thousand neurons of the heart rival the subcortical centers in the brain, and the heart sends more information to the brain than the brain sends to the heart. The heart has its own magnetic field, which is five thousand times more powerful than the brain and produces 40-60 times more electrical power.

Surprisingly, the lining of the intestines also contain nerve cells with their own form of intelligence. When we say we 'know in our heart' or have a 'gut feeling' our assessment is accurate. While the heart and gut produce an intelligence that's more intuitive and direct than the brain's logic, it's no less valuable. In fact, Doc Childre and Howard Martin of the Institute of HeartMath explain that the heart's response is more selective than the brain's. We've been taught that the brain is the control center of the body, but now scientists understand that while the heart chooses whether or not it will listen to the brain, the brain always obeys the heart. As we said earlier, the brain prefers the safety of the status quo, even when stasis is not our best option. But as Childre and Martin point out, "The heart isn't only *open* to possibilities, it actively scans for them. The intelligence of the heart acts as an impetus for what some scientists call *qualia*—our experience of the feelings and qualities of love, compassion, nonjudgment, tolerance, patience and forgiveness."

By now, it should come as little surprise that sages understood heart intelligence long before science 'discovered' it. Rumi explained the connection between heart intelligence and the

One Mind when he said, "There is a rope of light between your heart and [Source] that nothing can weaken or break, and it is always in His Hands." The unmanaged brain and a neglected or hardened heart will keep us in thrall to the little self and focused on our material illusion. When we access the emotional intelligence of the heart, we find it to be the gateway to transcendence Rumi described. What masquerades as love in this world is an extremely pale and inaccurate shadow of the real thing. *Transcendent love has never been an emotion, an intellectual attribute or a commodity to be given, received or denied; it's an all-encompassing state of **Being** that knows no boundaries or conditions.* Let's take a look at a few words of wisdom on the subject of love:

Love is too hopelessly abused and inadequate a word for the state I experienced, but I have found no other—Joseph Chilton Pearce

Divine laws are simpler than human ones—which is why it can take a lifetime to be able to understand them. Only Love understands—Rumi

In one's search for truth, the first lesson and the last is love—Hazrat Inayat Khan

The ferry to any shore, to any land, to any realm, it is the wine cup, the heart—Hafiz

Divine love cannot deny its very Self. Divine love will be eternally true to its own being, and its being is giving all it can—Meister Eckhart

I [Source] am the Self in the heart of every creature—*Bhagavad Gita*

The Lord of Love dwells in the hearts of all. To realize him is to go beyond death—*Taittiriya Upanishad*

The Self is hidden in the lotus of the heart—*Chandogya Upanishad*

Bright but hidden, the Self dwells in the heart—*Mundaka Upanishad*

The Self, pure awareness, shines as the light within the heart—*Birhadaranyaka Upanishad*

This world believes that hate is the opposite of love, but from a spiritual perspective the opposite of love in this world would have to be fear. Why? Because fear is the main barrier that stands in the way of our becoming love. The introduction to *A Course in Miracles* makes an important point when it says, "The opposite of love is fear, but what is all-encompassing can have no opposite." In a world built on duality, opposites are extremely important to us, but Source embraces, balances, and therefore transcends, duality. In Reality, love can have no opposite. As *A Course in Miracles* goes on to explain, "Perfect love casts out fear, if fear exists, then there is not perfect love. But: only perfect love exists. If there is fear, it produces a state that does not exist." This may sound confusing at first because we may be confused by the word 'perfect.' Religion often teaches us that we have to be perfect (without sin) while at the same time telling us how sinful and imperfect we are. But that is not what the *Course* is saying. It is Divine love that's perfect, and that love perfects the love of anyone willing to let go of duality and BE love. When we're in fear, we're living from the little self. When we *are* love, we're living from Self and fear can no longer exist. Source doesn't demand that we perfect our love, only that we be willing to let go of anything in us that is not love. Our willingness itself is enough of an invitation for Self/Love to support, assist and direct us.

It should come as little surprise that your willingness to be love will evoke some deep-seated fears held by the little self. Many people have told us their greatest fear is what they will find inside themselves. Others fear that love will make them too vulnerable to exist in this world, or the world will reject them when they no longer fit in with its agenda. And the little self will try it's best to convince you that you will lose yourself and be swallowed up by Source. The number of fears the little self has invented would probably outnumber the grains of sand on all the world's beaches, but Divine Love answers each one of them. As Rumi promised, "Love is the Bridge between you and everything." Once again, this can't be learned but must be experienced. As the 17th century mystical writer Frances de Sales explained, "You learn to love...by loving. All those who think to learn in any other way deceive themselves." However, you don't have to jump in with two feet. Becoming love usually involves baby steps that combine letting go of anything in you that is not love, while learning to trust that Divine love is safe. Hafiz spoke from his own direct experience when he said, "Your fidelity to love, that is all you need. What was once a fear or problem will see you coming and step aside, or run."

Part Four

Part Five

Living Fearless Spirituality

I. Trust

No matter what religion or society conditions us to believe, no one else can save us. Only in duality can we conceive of 'levels' of power and the inequality needed to allow for the gap between a 'savior' and those who cannot save themselves. In the oneness of our quantum Reality, we all share equal power and equal responsibility for the use of that power. We are all using our creative power to project this world, which is an accurate reflection of our collective core desires. In our conscious universe, free will and choice are paramount. Since we chose to create the dualistic world we find ourselves in, we must also choose to escape it. The bottom line: you are your own savior/ superhero. Whether you accept that role or reject it, this truth remains. Rejecting the role means the illusion of the dualistic birth/death cycle will continue. Accepting it will release you from the cycle but involves something the little self has trained us *not* to do: trust our own inner knowing! Yes, this means letting go of everything that constitutes our outer defense and support systems; all those things that keep us from connecting with Self.

Plotinus explained, "For that which we seek to behold is that which gives us light...How can this come to us? Strip yourself of everything." Krishnamurti advised, "To be a light to yourself you must deny every other light, however great that light be." If a hundred candles are burning in a room, it's impossible to recognize the light coming from one. But the instant ninety-nine are blown out, the source and light of the hundredth candle becomes evident. Of course that inner light we want to reveal shines from the Self. Like any light, it's impossible to cover it up and still see it. When the little self is in charge the light of Self is veiled. Each time we hand out power over to others, a veil is thrown over the light. If we continue to believe that other lights shine brighter than our own, polarized duality will inevitably demand it's due. Why? *When we put our trust in someone or something outside us, we make that person or thing into an idol and ourselves into a slave!* When that happens, we've given someone else the power to control and victimize us.

In Reality, since everything is one thing, unequal access to personal power cannot exist, thus victimizers and victims can't exist. However in virtual reality, our belief in levels of unequal power makes it real for us and we experience the consequences of those beliefs. As Kahlil Gibran realized, "Beyond this burdened self lives my freer Self; and to him, my dreams are a battle fought in twilight and my desires the rattling of bones." Those 'dream battles' will go on and on, as long as we compete for specialness. It is only by shifting our thinking back to Self, back to what is Real, that we can let go of victim/victimizer thinking and realize that we've had the power of choice all along. So, how do we learn to let go of the little self and learn to trust Self?

Since spiritual awakening is an 'experience-it-yourself' project, learning to trust the Self is essential for the process to succeed. And yes, trust is a process, not a light switch we can flip. It takes baby steps, tiny moments of listening for the inner voice and responding. As we test that voice and find that it never lets us down, we're able to take bigger steps. Don't be surprised if what you hear appears to be vastly different than what you might have expected to hear. In fact, this is a good time to let go of all expectations except one, the expectation that you will succeed. What you hear may not fit at all with the brain's logic, but remember, Self is not a slave to the limited perception of the brain and senses. It has access to the Divine Mind and 'knows' what your highest good is. This 'good' is not directed toward a more comfortable life in illusion, but is always focused on your return to Reality. Admitting that the brain and little self know nothing in comparison to the Self makes this adjustment easier. Just ask the little self what it has done in your best interests lately (or ever for that matter). Wait for an answer, which you will never get, and you'll know where to put your trust. Eventually, you will come to the point where you must let go of the little self's chatter and the brain's logic; and rely fully on the guidance of Self and Source. Don't fear that you're giving away your power or losing yourself in the Divine. *Instead, you're finding yourself; you're recognizing and reclaiming the fullness of power and understanding that have always been yours.*

Since the continued existence of the little self depends on our desire for separation and specialness, it does everything possible to keep us from hearing the voice of the Self. The little self is like a magician that gets you to look in the opposite direction while they slip something past you. One of its most successful strategies is our conditioned belief that the brain's logic is absolutely essential to the body's survival. For the brain to reason, to construct logical arguments, it must have information. Most of us accept the daily onslaught of information that inundates our world, but very few question whether all that information is actually necessary. Instead, we assume that information will give us the edge we need to secure our safety and 'get ahead' in the world. Even when the information backfires, we look for more information to replace it. On a daily basis we're confronted with constantly changing information produced by 'experts' who rarely agree on anything, and regularly change their minds. When they do, they act as if we should have known better than to believe the ideas they sold to us yesterday or completely disavow ever having said it. Is it any wonder that so many feel confused, frustrated, insecure and fearful?

We're so addicted to information; we entertain ourselves with an endless stream of titillating, yet useless, information about the goings on of the 'rich and famous.' But no matter how much information we devour, this world's data will always be based on flawed perception and guesswork. Nonetheless, it keeps us so involved with the brain's valueless chatter; we can't recognize the treasure within. In fact, the overabundance of information has convinced us that there's just too much for us to know, so we have no other choice but to rely on experts to filter it for us. This glut of and contradictory information and advice has caused most of us to either become fearfully dependent on one expert after another or become equally fearful of trusting anyone or anything, especially ourselves. But we have reason to take heart. As Eknath Easwaran observed, "The seers discovered a core of consciousness beyond time and change...There is an infinite changeless reality beneath the world of change...discover this reality experientially."

Since the Self won't violate our free will by speaking up unless invited to do so, the little self appears to have the advantage. The voice we need to hear is always there, but we can't invite it to speak if we're not even aware of its existence. That's one of the main reasons sages speak about the value of silence. Shankara explains, "Silence is a state of entire peace in which the intellect ceases to occupy itself with the unreal." To become silent, we can begin by ignoring much of the valueless information the world throws at us. With less information, the brain has less to fuel its useless chatter. When we take the time to pay attention to the chatter, we realize how worthless most of it is and we can choose to detach from it. That doesn't necessarily mean the brain is going to be quiet, but we can become like the person who lives by the railroad track but is no longer aware of the noise made by the trains rushing by. We hear the noise when we're attached to it. The louder it is, the more attached we are.

Tuning out the senses and the world's information appears to be in opposition to Jesus' words recorded at Matthew 7: 7-8, "Ask, and it will be given you; seek, and you will find; knock and it will be opened to you. For everyone who asks receives, and he who seeks finds, and to him who knocks it will be opened." Most take this saying at face value, so they fail to ask some pertinent questions such as, who do we ask and what do we ask for? What do we seek, and where do we knock? Most religions' favored form of asking is prayer. Even though prayer may be silent, it's directed to a being outside ourselves and usually consists of talking rather than listening. Instead of letting go of preconceived notions and willingly hearing what the Divine wants to tell us, prayer often devolves into a ritualized form of begging that focuses on the things we believe will improve the quality of this life. What should we seek? Again, many seek for the things of this world while others look for mercy and forgiveness of sins. Where do most religions tell us we should knock? Evidently on the 'door of heaven' that supposedly exists outside us and stands between us and all that we might desire after the death of the body. Many are convinced that obedience, faith and works are the keys that will open the door. Although Jesus appears to be telling us to look outside ourselves, let's get a more complete picture of what he meant.

Keep in mind that followers who don't experience for themselves have no way to completely assimilate what a master is saying. And unfortunately, their own attachments and aversions filter everything they hear. The writer of the Bible gospel quoted above recorded what he *believed* Jesus was saying, while Jesus' gnostic followers understood his words in a very different way. In the gnostic writing *Dialogue of the Savior,* Jesus says, "Light the lamp within you...Knock on yourself as upon a door and walk upon yourself as on a straight road." And the Bible also omits these words found in the gnostic *Gospel of Thomas*, "Jesus said, 'If you bring forth what is within you, what you have will save you.'" As often is the case, truth appears to create a paradox. At first glance, we might take the asking, seeking and knocking Jesus spoke of literally, and decide they require physical action that involves the senses and takes place outside us. However, Jesus himself, as well as hundreds of other sages, makes it clear that if we keep looking outside, if we *do* rather than *be*, we'll continue to be disappointed.

Like Jesus, Nisargadatta knew from experience that, "Your own Self is your ultimate teacher…It is only your inner teacher that will walk with you to the goal, for Self is the goal." When Jesus said, "Ask, and it will be given you," he was telling you to ask Self. When he said, "…seek, and you will find," he meant that the seeking would be done within. When he said, "…knock and it will be opened to you," he was telling you to knock on the inner door that only the Self will open. Lao Tzu made a similar observation when he said, "Without going outside you may know the whole world. Without looking through the window, you may know the ways of heaven. The master observes the world but trusts his inner vision. He allows things to come and go. His heart is open to the sky." But how do we find the inner door? How can we knock on something inside? Answering the question, Rumi tells us, "Knock on the inner door. No other. Why are you knocking at every door? Go, *knock at the door of your own heart.*" [Italics ours]

Allowing the little self to shift our trust toward someone or something outside us has robbed us of the ability to realize that all asking, seeking and knocking is internal. This shift has caused us to forget that the Self already has the answers to all our questions, answers that are based in truth rather than illusion. How do we return to the full consciousness of Self? As Rumi pointed out, we knock at the door of the heart. That may sound odd, but remember that sages often use symbolic language to make a point. We know there's no literal door in the heart, but in this case the door stands for the 'door' we shut to divide off a small portion of consciousness. A choice must be made to open up our access to the Divine Mind. Visualize yourself making a surprise visit on a friend that you haven't seen in years. You are unaware that they're napping in a favorite comfy chair. If you just stand quietly outside the door, they won't realize you're there. But you're not just going to walk away because they don't sense your presence; you're going to knock, ring the bell or call out to them. Once they wake enough to realize someone is at the door, they wipe the sleep from their eyes and come to answer. Similarly, we've put our heart connection to Self into a sleeping state while the little self is in charge. It's our 'asking, seeking and knocking,' that demonstrates our willingness to reopen that connection. If you've thought of the heart as a mere circulatory pump this might sound ridiculous, but as we discussed in Part Four, science has finally recognized the heart is far more than that. As sages have always known, the heart is literally your lifeline, your unbreakable connection with the Divine.

It can be difficult for us to recognize the heart's value since the little self has worked so diligently to cheapen it. How? By claiming the heart is the source of sticky sweet sentimentality, unstable emotional behavior, uncontrolled passions and treachery. Tune into almost any radio station, and within a few moments you'll hear a song that reinforces our fear of the heart and/or misrepresents what love actually is. Religion adds to the problem by convincing us the heart is not to be trusted. The *Bible* book of Matthew 6: 19-20 states, "Out of the heart come evil thoughts, murder, adultery, fornication, theft, false witness, slander. These are what defile a man." It's true that the desires behind our emotions originate with the little self; however, the brain clings to these desires and as science is now demonstrating, the entire body is capable of responding with emotion. And in a vicious cycle, thoughts and emotions are reinforced by the addictive chemicals the body produces which cause us to repeat the same valueless thoughts and emotions over and over again. Instead of believing the heart can't be trusted, we can benefit by learning to distinguish the difference between valueless thoughts and emotions and valuable emotional intelligence that comes to us from Self through the heart connection. Rumi realized, "As you live deeper in the heart, the mirror gets clearer and cleaner." And the 20th century sage Ramana Maharshi clearly stated, "To identify oneself with the body and yet seek happiness is like attempting to cross a river on the back of an alligator... You are the abiding Reality, while the intellect is just a phenomenon...The heart is the only Reality. The mind is only a transient phase... To remain as one's Self is to enter the Heart." It is through the heart that the bridge to the One Mind opens, and this is the pathway of gnosis.

II. Gnosis and the Heart

Love means...to open the eyes of inner vision—
Rumi

Over 1,700 years ago, the philosopher Plotinus shared this powerful observation: "This is not a journey for the feet; the feet bring us only from land to land; nor need you think of coach or ship to carry you away; all this order of things you must set aside and refuse to see: you must close the eyes and call instead upon another vision which is to be waked within you, a vision, the birthright of all, which few turn to see." Although Plotinus argued with religionists over the use and meaning of the word 'gnostic,' it's obvious in the quote that he was referring to gnosis when he explained that we must "close the eyes and call instead upon another vision which is to be waked within you." Rumi shared a beautiful description of gnosis when he said: "Make everything in you an ear, each atom of your being, and you will hear at every moment what the Source is whispering to you, just to you and for you, without any need for my words or anyone else's. You are—we all are—the beloved of the Beloved, and in every moment, in every event of your life, the Beloved is whispering to you exactly what you need to hear and know. Who can ever explain this miracle?

It simply is. Listen and you will discover it in every passing moment. Listen, and your whole life will become a conversation in thought and act between you and Him, directly, wordlessly, now and always. It was to enjoy this conversation that you and I were created."

When Rumi suggested that we turn every atom of our being into an 'ear' it was a polite, poetic way of telling us to shut up and make listening to our inner voice our absolute priority. But the heart can also take us past the limited sight of the body to the infinite 'vision' offered by the heart. Antoine De Saint Exupery, a 20th century aviator and writer correctly pointed this out in his book, *The Little Prince*, "It is only the heart that can see rightly; what is essential is invisible to the eye." The 10th century Sufi mystic Mansur Al-Hallaj described the heart's part in gnosis when he said, "I saw my lord with the eye of my heart and I said, 'Who art thou?' And he said, 'Thou.'" We could fill countless pages with quotes on the essential part the heart plays in reconnecting us to the One Mind of the Divine; here is just a small sampling:

The Eye of my heart sees everything—Black Elk

The door of the heart must be opened—Shams-iTabrizi, 12th century mystic wanderer and spiritual companion of Rumi

Lift the veil that obscures the heart and there you will find what you are looking for—Kabir, 16th century mystic poet of India

Mind creates the abyss, the heart crosses it—Nisargadatta

One who knows the distances out to the outermost star is astonished when he discovers the magnificent space in the heart...Work of the eyes is done; now go and do heart-work—Ranier Marie Rilke, 19th century mystic poet

The heart outstrips the clumsy senses, and sees...an undistorted world—Evelyn Underhill, 20th century Christian mystic, author of *Mysticism.* She was often criticized for saying a mystical connection to the Divine is open to all.

Idols...are our beliefs, our cherished preconceptions of the truth which block the unreserved opening of...heart to Reality—Alan Watts

The awakened heart is like a luminous sphere—just giving without thought to any who may come close—Meister Eckhart

One of the main functions of formalized religion is to protect people against a direct experience of God. Who looks outside dreams. Who looks inside awakes...Your vision will become clear only when you look into your own heart...The knowledge of the heart is in no book and is not to be found in the mouth of any teacher—Carl Jung

Self is what you are. You are that Fathomlessness in which experience and concepts appear. Self is the moment that has no coming or going. It is the Heart...It shines to Itself, by Itself, in Itself. Self is what gives breath to Life. You need not search for It, It is Here. You are That through which you would search. You are what you are looking for! And That is All it is. Only Self is—Papaji

There is probably no other sage who spoke so fluently of the heart's role than Rumi. Earlier in the book, a quote from Rumi made it clear that our escape route from virtual reality is through the heart. Through direct experience he realized, "There is a rope of light between your heart and His that nothing can weaken or break, and it is always in His hands." Here are a few more of Rumi's observations concerning the important part the heart plays:

The heart has its own language

I looked in temples, churches, and mosques, but I found the Divine within my heart.

The only scripture I recommend is your own spiritual heart.

Stop the flow of your words, open the window of your Heart and let the Spirit speak.

Your heart is the size of an ocean. Go find yourself it its hidden depths.

Be awake always, be a seeker of the heart.

Close the door of words that the window of the heart may open.

Everyone sees the unseen in proportion to the clarity of the heart.

He who keeps his heart awake...While the eyes of his head may sleep, his heart will open hundreds of eye

If your heart is already awakened, sleep peacefully; sleep in the arms of Love, for your spiritual eye is not absent

The heart's the essence, words only the accident. The accident's accessory, the essence is what matters

I burnish bright the mirror of my heart until at last, reflected for my rapture, the Self's eternal beauty appears.

III. Belief, Faith or Trust?

But how do we go about hearing or seeing without engaging the senses? How can the heart speak and see? So far Part Five has emphasized the importance of looking past sense perception and shutting out the world's so-called 'wisdom' that only comes to us through the senses. We want to make it very clear that tuning out what the senses offer doesn't mean you'll need to seclude yourself in an ashram or monastery; becoming a monk or a recluse or be so detached from the world no one wants to be around you. The ancient *Isha Upanishad* is clear on this point, "Renunciation is renunciation of the self—not of life. The end...is to know the Self...to realize your identity." If spirituality is not about renouncing the world or literally separating ourselves from it, what should we do? Eknath Easwaran made an excellent point when he said, "We start wherever we are, not running away from society, but right in the midst of life."

Often those who literally separate themselves from the world find that they cannot sustain their spirituality when they do have to interact with it. When we talk about tuning out the world's 'wisdom,' we're talking about an attitude change. *We can hear and see what the world has to offer, and we can continue to be involved in life, but we are no longer invested in the misperception that it's real or valuable.* This is what Jesus meant when he told his follower that like him, they were "in the world" but not "of the world." (John 17: 16) This may sound difficult, but as we divest ourselves of conditioning and commune with the One Mind, the transition feels right and peaceful. It's like wearing an invisible raincoat; you're in the world's storm, but you're dry and comfortable, insulated from the weather.

Instead of thinking of our inner voice as an intuitive feeling or a conscience that pinches us when we're about to violate our own standards, it's the voice of Self in direct connection with the One Mind of the Divine. In this world we communicate using a variety of means, but the most common methods involve spoken and written language. Language can be extremely powerful, but one of the major problems of language is its fluid nature.

Communication breaks down because each 'speaker' has their own meaning for the words they use, a meaning that may not be shared by the listener. Communications experts call this disconnect 'noise in the channel,' and it happens far more often than we might realize. Anyone who has gotten into an argument with someone over misunderstood wording has experienced this noise. Communication not only depends on the clarity of the speaker, but the attentiveness of the listener. Remember that the brain is often busy thinking of arguments or replies rather than paying close attention to what is being said. When we add the complication of 'body language,' we can see how the body's means of communication can be confusing. But unlike the little self, the One Mind is not limited by human means of communication.

Hermes Trismegistus, the purported author of the *Hermetic Corpus*, a series of works encouraging direct communion with the Divine wrote, "Understand that what sees and hears inside you is...The Mind of God." This Self/One Mind/higher consciousness bypasses the brain and senses and doesn't bother with our fallible forms of language. Instead, the One Mind communes with us through the intelligent heart where language is unnecessary. No, we're not just talking about emotions either. We may use terms like 'inner voice' or say that Self speaks to us, but we don't so much 'hear' that voice as experience it. Instead of taking in information and assimilating it bit by bit as the brain does, the One Mind presents us with what we've come to think of as a 'thought bundle.' You can't know exactly what this is like until you've experienced it directly, but it could be described as a 'knowing,' a rich and complete concept that's delivered and understood instantly. What we're describing is not at all like a drug or meditation induced 'vision' although there may be a visual component. Rather, it's an immediate assimilation of truth that takes us from the explicate to the implicate, from virtual reality to Reality.

This experience can feel quite shocking at first since even our wildest imaginings can't prepare us for Reality. Earlier we looked at what Jesus had to say about it in the gnostic *Gospel of Thomas*: In the Greek version, "When one finds, one will be astonished," and the Coptic version substitutes the word "troubled," but both words fit. Yes, it's probable that we'll be both astonished and troubled when we experience All That Is. Why? Because Reality is not only astounding, it's truth destroys the world we assumed was real. When a direct experience of the Divine appears to have little, if anything, to do with logic; it is the logic that's deficient. As Adyashanti put it, "Enlightenment is the crumbling away of untruth. It's seeing through the facade of pretense...Enlightenment is...the complete eradication of everything we imagined to be true." How do we trust such an experience; how do we know we've opened the door to the One Mind and haven't just lost our mind? To put it bluntly, you will 'lose your mind' but what we mean by that is the conditioned mind/brain controlled by the treacherous little self. This is no loss since the little self has always put its existence ahead of any body it was projecting. Instead, you'll be guided by the Self, the real you, that cares only about your best and highest interests. Regardless, you may find it difficult to trust the Self and accept what you're being shown. Trust may be easier to understand and feel comfortable with when you learn how it differs from belief.

Most religions, and many spiritual teachers, talk about faith and trust, but what they're actually describing is belief. Although the thesaurus considers belief, faith and trust to be interchangeable synonyms, from a spiritual standpoint, belief bears no resemblance whatever to either faith or trust. In many aspects of life, we're conditioned to believe, usually by authority figures that have something to gain from our belief. But what does belief actually mean? *Belief is the conviction that something outside us is true even though we have* **no proof** *that it is true.* Instead of originating with our own direct experience, belief is based on faulty, limited perception. And to make matters worse, it's almost always someone else's perception and/or opinion. As we've learned, perception means that objective reality (truth) cannot exist in this world. Religious and spiritual beliefs are not only based on misperception, they have usually been passed around for centuries and gone through hundreds, if not thousands, of people before they reach you. Frankly, belief is little more than an illusion a believer chooses to cling to, usually because another equally ill-informed believer persuaded them to, whether kindly or forcefully. A large percentage of believers don't understand and can't explain what they profess to believe. A recent poll revealed that 90% of Christians have not read the complete Bible, yet they say they firmly believe what it says.

Belief is often based on a static set of doctrines compiled by people the believer does not know and has little or no legitimate reason to trust. Accepting a belief system simply because our grandparents or parents did has little merit. For example, a young person was about to bake a ham. Before putting it in the oven, they cut off the end the ham. When a friend asked them why they had done that, they replied that was how their mother did it. Since the young cook believed that their mother knew best in the kitchen, they had accepted the practice without question. Curious about the reasons for doing this, the young person ended up calling their mother who explained it was how her mother did things, and she, in turn, believed in her mother's cooking ability. Since they were both baffled, they called the grandmother who explained that her roasting pan was small and an entire ham would not fit unless she cut off the end. How many beliefs imprison us that have no more value than that?

Beliefs are stagnant because they encourage the brain to tune out or shut down when confronted with thoughts that differ with the accepted belief system. James P. Carse defines belief this way, "Belief is a corrupted form of knowledge that refuses correction...A belief system is effective only if it can place itself in opposition to another...the more threating the opposition the better." Once a believer has decided their accepted 'thought system' is true, the brain automatically begins to formulate arguments in its defense. The brain also labels anything that opposes the belief system as untrue and readies arguments against them. Like the bucket we spoke of earlier, the brain filled with beliefs has no room for new ideas. Remember too that the brain loves the status quo, so beliefs often become among its most cherished security blankets. Alan Watts explained, "Belief...is the insistence that the truth is what one would...wish it to be...If we are open only to discoveries which will accord with what we know already, we may as well stay shut... Idols...are our beliefs, our cherished preconceptions of the truth which block the unreserved opening of mind and heart to Reality." (The idea that 'disbelievers' can exist is rather ridiculous since they staunchly believe 'against' rather than 'for.' 'Disbelievers' often try to masquerade as skeptics, but a genuine skeptic holds an open mind and is not swayed one way or the other until they're convinced by facts.)

Belief systems are certainly not limited to religion, and the results can be disastrous when applied to any area of our life. We can just as easily develop unshakable beliefs about our country, economic system, political party, medical treatment, nutritional or exercise program etc. Scientists often become as adamant as religionists when they turn their theories into literalist doctrines and worship their supposed infallibility. Although quantum physics demonstrates that consciousness is the foundation of the universe and permeates everything in existence, many scientists (material realists) tenaciously cling to the belief that consciousness evolved from matter and is nothing but more matter. Regardless of the fact that quantum physics has continually been proven correct and material realists have been unable to find even a shred of evidence to support their postulate, this belief has become their credo, passed down from one generation of scientists to the next. It may give us a feeling of comfort to accept a belief system that's been around for hundreds of years and/or been accepted by thousands of people, but time or numbers can't make anything true. When we insist on clinging to illusory comfort we pay a terrible price: knowing truth firsthand. Most of us have been content to keep projecting illusions for countless lifetimes, but when we're ready for Reality, illusion and belief will no longer do. Then you're ready to step into faith/ trust based on direct experience.

Let's examine exactly why faith and trust are so different from belief. The dictionary defines both faith and trust as "firm reliance on the integrity, ability, or character of a person or thing...a feeling of certainty that a person or thing will not fail...a confident expectation of something." Spiritually speaking, neither faith nor trust gives us a "feeling of certainty" and "confident expectation" simply because we want them to. Instead, we feel certain and confident based on our own direct, personal experience. As we've learned, truth can only exist in the implicate portion of the universe since the explicate portion is a projection based on duality and filtered through perception. The bottom line: we can only know truth by reconnecting with consciousness at the implicate level. Virtual reality/the explicate order is a place of ever-changing illusory 'facts.' Reality/the implicate order is the realm of unchanging truths. *Truth is absolutely reliable because it is what IS: the unchangeable Divine.* Trust/faith helps us to supports our understanding of truth because they are fluid and dynamic rather than static; trust allows us to expand from illumination to illumination. In other words, trusting Source and Self is expansive. It's not a blind trust, nor does it demand a 'leap' of faith that has no foundation. While belief requires us to accept what we haven't experienced and don't understand, gnosis builds trust exactly because it *is* experience. The more we experience, the more we can see that we have legitimate reason to trust.

When you learned to ride a bike or roller skate, your trust/faith in your ability grew each time you built on your experience. At first you may have been quite fearful of falling, but eventually, the thought that you ever mistrusted your ability seemed ridiculous. But let's not forget that while you were in the process, you did occasionally fall. Children often proudly display the scabs and bruises they acquired while experiencing something new. But as we get older, our tendency is to avoid experiences that involve possible pain or failure. We would be deceiving you if we said there was no pain involved in the transition from self to Self. As Carl Jung learned through experience, "There is no coming to consciousness without pain. People will do anything, no matter how absurd, in order to avoid facing their own soul. One does not become enlightened by imagining figures of light, but by making the darkness conscious." But keep in mind, the hurt involved is always self-inflicted, never from by Source. Nonetheless, it can still feel painful when the light of truth shines full force on our conditioning, attachments and aversions. Think of it as a dirty bandage over a festering sore. Yes, it will hurt to pull the sticky bandage off, but the sore won't heal until we have the courage to feel the pain and pull it off. However, the severity of that pain depends on how tightly we cling to the valueless. However, the word 'failure' is strictly another illusory concept manufactured and owned by the little self. As far as Source is concerned, our willingness to experience truth (even if it's just our willingness to be willing) guarantees our return.

Rabia Basri, an 8th century Sufi mystic was able to proclaim, "The one who tastes, knows," because she experienced for herself. At some point we have do as Rumi advised, "Leave the rind and descend into the pith." However, the little self can easily convince us that the rind alone will suffice. It's only when we begin to let go of the false self that we truly understand the difference, and therein lays the difficulty. Aristotle cautioned, "We are what we repeatedly do." We can't experience something new if we continue to think and do what we've always thought and done. Much as we might like to think otherwise, we all know that we can't lose weight until we change our eating habits and start exercising. The same is true spiritually. Until we choose to stop consuming the world's 'junk food' information and second hand beliefs, we won't experience Source. Nonetheless, letting go of the little self while we still lack trust in Self and Source can *feel* terrifying. It's actually an exhilarating and liberating thing to do, but the little self will be making a last ditch effort to protect its own existence. Even when we begin to realize the little self is behind the world's suffering; it's a misery that we're accustomed to. Like the person who clings to an abusive relationship or gets into one hurtful relationship after another, most would rather cling to the false self and let it keep beating them into submission rather than experience something new. As Thich Nhat Hanh, a 20th century Zen Buddhist monk observed, "People have a hard time letting go of their suffering. Out of a fear of the unknown, they prefer suffering that is familiar." That's why trust is a process, not a leap.

Trust is a component of love and for love to be real, free will must be involved. Source *is* love and love knows that *anything given or taken by force, coercion or with reservation or restriction cannot be love.* Imagine that you hadn't seen your mother or father since you were an infant. As you were growing up, you heard many second and third hand stories about them, some positive, some negative. When you finally meet, would you be able to instantly turn on your trust and love like a light switch? Of course not. Although you might feel drawn to your absent parent, you would expect that love and trust would grow slowly, over time, as you experienced what he or she was like in many different situations. Although we are the stuff of Source and we are the Self, we've forgotten that connection and need to take the time to become reacquainted. As Rumi said, "Listen, and your whole life will *become* a conversation in thought and act between you and Him, directly, wordlessly, now and always." [italics ours] The operative word here is *become*.

Although our emotions and the brain's logic often let us down, Self never will. How can we be so certain of this? Unlike the little self, Self constantly communes with the One Mind. And like the One Mind, it can literally 'see around corners, something sense perception and the brain can never do. In *Dialogue of the Savior* the 2nd century gnostic writer Silvanus said, "Bring in your guide and your teacher. The mind is the guide...Acquire strength, for the mind is strong...enlighten your mind...Light the lamp within you." As we now know, Silvanus was speaking of the level of consciousness of the Self and One Mind, not the brain or little self. When he says, "enlighten your mind...Light the lamp within you," he's referring to the 'light energy' that, along with consciousness, is the matrix of all that is. The early gnostic writing, *Creation of the World and the Alien Man* explains, "There is no boundary for the light and it was not known when it came into being. Nothing was when light was not, nothing was when radiance was not. Nothing was when the Mighty Life was not, there never was a boundary for the light." As you've probably guessed, this light is far more than a poetic symbol.

Scientists have discovered that tiny particles of light, called biophotons, are emitted by all living things. Plainly stated, everything is life, and all life is light. In the *Gospel of Thomas*, Jesus took this description a step further when he recognized the Divine quality of that light. He said, "If they say to you, 'Where have you come from?' Say to them, 'We have come from the light, from the place where the light came into being by itself, established itself, and became manifest through their image.' If they say to you, 'Is it you?' say, 'We are its children.'" When we 'bring in the light,' we are shedding our connection with virtual reality and reclaiming our truth as beings of light, inseparable from the Divine. Unfortunately, we rarely recognize this Divine light in ourselves or others. When Jesus said, "For this reason I say, if one is whole, one will be filled with light, but if one is divided, one will be filled with darkness," he wasn't using light as a symbol of good or darkness of a symbol of evil. Instead he meant that our desire for separation had blinded us to the literal light of our true nature.

The gnostic *Dialogue of the Savior* agrees saying, "As long as your hearts are dark, your light...is far from you." That's why Jesus told his followers, "No one lights a lamp and puts it under a basket, nor does one put it in a hidden place. Rather, one puts it on a stand so that all who come and go will see its light." Most Christian churches believe that these words were meant to encourage Jesus' followers to preach about him and/or to use their 'God-given' talents for good. But when we realize that we are literally beings of light, we see that he was telling us to quit identifying with the bushel basket (body) that hides our true 'light' identity. No matter what we might achieve as the little self, it has absolutely no value compared to this recognition. That may sound harsh, but all 'saving' of the body that's done in this world leaves the one who is saved trapped in the cycle of birth and death! The Bible says that Jesus fed the hungry, cured the sick and even raised the dead, but the same people that he helped were hungry and sick again and their body eventually died. Because these were all temporary fixes, Jesus told his listeners they would, "...do greater works than these." (John 14:12) When it came to 'waking up' Jesus knew he could not be a savior to anyone except himself, just as you must save yourself. This "greater work" involves no physical action. It is the internal work we do to symbolically 'feed, heal and resurrect' our connection with Self.

Even so, when we reconnect with Self, the Divine light we radiate in this world can be no more than a reflection, not the light itself. Why? Because physical senses are limited to perception; and perception will never be able to reveal our implicate Reality. Regardless, there is nothing finer that we can do in this world than reflect the light of Divine love. As Rumi reminds us, "You think you are earthly beings, but you have been kneaded from the Light of Certainty. You are the guardians of God's Light, so come, return to the root of the root of your own Self."

IV. Stay or Return?

Sooner or later each of us will make the ultimate choice: do I want to continue projecting an illusion of perpetual fear, or embrace the spiritual fearlessness that comes from understanding that this world is nothing to fear? Papaji made an excellent point when he said, "If these attachments give you happiness and peace of mind then stay with them because it isn't time to leave them." It is possible to look within and find that we're not ready to let go of the separation and specialness. If that's the case, we can continue to play in this world. However, Papaji asks that we also make certain we're not just sticking with our attachments because we're used to them and afraid to change. He added, "But if you see [attachments] for what they are—false, empty promises—then it is time to reject them. It is no use to experience what has already been experienced. If you know the fire burns there is no need to be burned again. Like this, avoid attachments like fire because they will burn you. Have a firm decision that you do not want to suffer and that you are here to realize Real Happiness in this incarnation. This is the most important decision."

The contemporary sage Anthony Paul Moo-Young, known as Mooji, recognized that even when we are ready to let go, the little self is not. He cautioned, "If you pay attention to the mind, there will never seem to be an appropriate time to turn towards your own Self...The mind will keep you busy, you'll find more and more things to do, more and more things that need to be attended to, and this is merely a trick." Whether we're aware of it or not, each seemingly insignificant choice we make in this world either strengthens or weakens the little self, takes us either closer to, or further away from, the Self. We make hundreds of choices each day that favor separation or Oneness, self-interest or love. As we make these choices, we either raise the volume of our inner knowing, or we lower it.

The 8th century Buddhist sage Padmasambhava (also known as Rinpoche or Padum), is considered by some to be a second Buddha. Among the writings attributed to him is the *Tibetan Book of the Great Liberation*, which was unknown in the West until the 1950s. His words are testimony to the fact that truth remains the same, whether it is spoken in the 8th century by a sage or in the 21st century by a physicist. The following are a few excerpts from Rinpoche's book:

"It is only because of deluded ideas, which you are free to accept or reject, that you wander in the world...Matter is derived from mind, and not mind from matter...All phenomena are your own ideas, self-conceived in the mind, like reflections in a mirror ...When you realize that all phenomena are as unstable as the air, they lose their power to fascinate and bind you...Your mind in its true state...is not realizable as a separate thing, but as the unity of all things...Until duality is transcended and at-one-ment realized, enlightenment cannot be attained...Although the clear light of Reality is possessed by all beings, it is not recognized by them...Without beginning or ending, your original wisdom has been shining forever like the sun...Again and again, look inside your own mind... It is surpassingly excellent...to know your own mind."

These excerpts demonstrate "there is nothing new under the sun." (Ecclesiastes 1:9) We've tried every possible political/cultural/

economic/religious combination and no matter how many times we repeat the pattern, it will continue to fail. As Papaji pointed out, once we know something burns, the wise course is to stop grasping it. As Rumi asked, "Why do you stay in prison when the door is so wide open?" If you've realized this world doesn't contain the answers you're looking for, it's time to let go of the little self. As we've said many times, there are no secrets, methods or practices necessary. Buddha observed, "Like swans leaving a lake, the thoughtful abandons one attachment after another." *No doubt all the sages we've quoted would agree what **is** necessary is the state of mind conducive to change, namely **willingness**.*

Once you're willing to let go of the little self (a state that can't be forced) there is no way for us to say what course your personal experience will take. Although everyone will eventually head back to Self and Source, you are starting in your own particular place. You have reached this place because the little self took you on a journey through countless lifetimes. Although there is no need for you to 'work your way back' through every experience, your particular journey will require its own route. Happily, willingness provides many shortcuts and Self gives sure directions. Although we can expect the little self to sneak in its belief in control and planning, this is a journey best left to unfold as it will, without judgment. Unlike the world's spirit of competition, there is no need to race for the finish line to claim the prize. You already have the prize and it's impossible for you to lose it. As Hafiz reminds us, "Mercy and Grace are just there; they are attributes of light, they want nothing but to be...God wants nothing in return for your existence. What madness to ever think you owe Him/Her/It anything... This is the time for you to compute the impossibility that there is anything but Grace."

V. Fear vs. Fearlessness

Even though this world is based on a dualistic concept that could not help but create fear, none of us planned on living in fear. We each believed that in separation we would be the special one; we would be the one who lived in complete wealth, power and security. If we hadn't held that belief very firmly, it would have been impossible to accept the enormous odds virtual reality stacks against us. Even when we do manage to claim a bit of specialness, we're never free from the fear of losing it or the desire for more. Rather than face the truth that it's impossible to beat the odds, the little self convinces us that we're the hapless victims of a cruel world. As a result, humanity spends vast amounts of time and money either running from fear or trying to get past it. There are countless ways we try to run from fear, but we primarily rely on whatever will dull or alter our mood. Whether it be food, shopping, gambling, drugs, drinking, video games, sex, entertainment, sports, excitement etc., etc., it's the 'running' aspect that usually leaves us addicted, exhausted and in greater fear than we were before.

Some try to cure their fear via professional therapy, self-help groups or do-it-yourself methods. While many of these programs do recognize the fact that fear exists primarily in our thoughts, they can't help but fall short. Why? Their 'cure' is based on the premise that managing our thoughts will eliminate our fears. That method can certainly give some relief from specific fears; but it's tantamount to taking a pill that treats our symptoms and does nothing to cure the disease itself. What good is convincing ourselves that it's safe to fly or drive when it's impossible for that belief to change the fact that planes and cars will still crash? *We get to the root of the problem only when we know that all fear is based on illusion.* We can spend time and money trying to run from or control our fears but *the most productive thing we can do is discover that there is no reason to fear.*

Quantum physics alone has discovered enough to determine that this world is an illusion. It has demonstrated that you are pure consciousness, not the material body projected by consciousness. Fear cannot exist in the implicate, quantum universe because it is whole. In oneness there is no 'other,' no opposition or competition. There is no scarcity since *everyone is everything, and we are all part of and one with the Divine.* As we've said earlier, in oneness we continue as unique beings that share a loving relationship with Source. Because the foundation of this relationship is love, its hallmarks are equality, peace, joy and cooperative creation that serves the highest good of all concerned. Even in virtual reality there have always been those who value such objectives and dream of creating just such a utopian system on earth. The earth has the ability to take care of everyone if we take care of it in return, but our dualistic thought system has been unable to sustain such a dream for long. No matter how well intentioned we may be to begin with, the false mind ensures that desires for separation and specialness interfere and before long fear will rule again. If that was not the case, humanity would have solved many of the world's major problems by now, yet we still struggle with inequality, poverty, disease, hunger and war. The only way to remove ourselves from fear is by removing ourselves from dualistic thinking. We begin to do this while we project virtual reality, and end it when we return to oneness.

Although the false self thrives on fear, it also suffers from the fear it foments. It uses fear to control, but in turn, it is also controlled by fear. How? The false self is like a person who has told so many lies they no longer have any idea what actually happened. For its own protection, the false self has tried to keep us in fear of the Divine. It has told us so often we cannot return to Source; it now believes its own lies. We can cling to this deception and live in fear, or we can join the sages and live a life of fearlessness. As the *Katha Upanishad* explained, "There are two selves, the apparent self and real Self. Of these it is the real Self...who must be felt as truly existing." Each of us can let go of the false mind and begin to reunite with Self and experience the Divine. As the 13th century philosopher priest, Thomas Aquinas accurately stated, "Truth never harms or frightens...God's compassion can never be limited; thus any god who could condemn is not a god at all...We use words like 'returning.' Inherent in that word is separation, and separation from God is never really possible." The anonymous authors who wrote *The Kybalion* based their words on gnosis, "Don't feel insecure or afraid; we are all held firmly in the Infinite Mind of the All, and there is naught to hurt us or for us to fear." You've heard the words of the sages and the scientists, now listen to the words coming through your own beautiful heart. The only security you'll ever need is the awareness of who and what you truly are. Knowing this, you have the key to fearless spirituality.

Move outside the tangle of fear thinking. Live in silence. Flow down and down in always widening rings of being—Rumi

Hope and fear are phantoms that arise from thinking of the self. When we don't see the self as Self, what do we have to fear?—Lao Tzu

Fear is the cheapest room in the house. I would like to see you living in better conditions... If one is afraid of losing anything, they have not looked into the Friend's eyes: they have forgotten God's promise... we are frightened every moment of our lives until we know the

Beloved... All the false notions of myself that once caused fear, pain, have turned to ash as I neared God—Hafiz

Part Five

Index

A

B

H

Notes

Notes

Notes

Notes

Thank you!

We appreciate the time you took to read

Fearless Spirituality

It takes a lot of courage to release the familiar and seemingly secure, to embrace the new. But there is no real security in what is no longer meaningful. There is more security in the adventurous and exciting, for in movement there is life and in change there is power—Alan Cohen

The important thing is this: to be able at any moment to sacrifice what we are for what we could become—Charles Du Bois

Fearlessness is already yours. It is our heartfelt desire that you allow nothing to stand in the way of experiencing this priceless gift.

We invite you to visit the Beginning of Fearlessness/Oroborus Books blog, website and free video eCourse. We look forward to your comments.

http://thebeginningoffearlessness.com